MEXICAN MACAWS

Scarlet Macaw. Painting by Barton Wright.

ANTHROPOLOGICAL PAPERS OF
THE UNIVERSITY OF ARIZONA
NUMBER 20

MEXICAN MACAWS

LYNDON L. HARGRAVE

Comparative Osteology and Survey of Remains From the Southwest

THE UNIVERSITY OF ARIZONA PRESS
TUCSON, ARIZONA 1970

THE UNIVERSITY OF ARIZONA PRESS

Manufactured in the U.S.A.

I. S. B. N.-0-8165-0212-9
L. C. No. 72-125168

PREFACE

Any contribution to ornithology and its application to prehistoric problems of avifauna as complex as the present study of macaws, presents many interrelated problems of analysis, visual and graphic presentation, and text form. A great many people have contributed generously of their time, knowledge, and skills toward completion of the present study. I wish here to express my gratitude for this invaluable aid and interest.

I am grateful to the National Park Service, U.S. Department of Interior, and its officials for providing the funds and facilities at the Southwest Archeological Center which have enabled me to carry on this project. In particular I wish to thank Chester A. Thomas, Director of the Center. I am further grateful to the Museum of Northern Arizona and Edward B. Danson, Director, for providing institutional sponsorship for the study as well as making available specimen material and information from their collections and records. The study was carried out under the provisions of Contract No. 14-10-3:931-2 between the National Park Service and the Museum of Northern Arizona. Barton Wright, Curator of Museum at the Museum of Northern Arizona, painted the representation of the Scarlet Macaw reproduced as the frontispiece.

A first requirement of this study was a collection of modern macaw material of known age, sex, and species. This need was filled through the kindness and generosity of three old friends: Bernard Roer of Roer Bird Farm, Phoenix, Arizona; Lewis D. Yaeger, Tepic, Nayarit, Mexico; and Robert W. Dickerman, Cornell University. They not only supplied specimens but drew on their own deep knowledge of the living birds to give vital information.

Hildegarde Howard, Los Angeles County Museum of Natural History, Alexander Wetmore and Richard L. Zusi of the Smithsonian Institution, Pierce Brodkorb of the University of Florida, and Richard S. Crossin and Stephen M. Russell of the University of Arizona generously made available additional modern macaw skeletons for use in the study.

The archaeological macaw specimens proved to be widely dispersed across the country in many institutions. Indeed, in many instances, there were doubts as to whether the specimens still existed. The following individuals undertook the search for these skeletal fragments in the collections of their respective institutions or in giving us leads to their location:

Dean Amadon, American Museum of Natural History (Pueblo Bonito: The Hyde Expedition Collections)

Douglas C. Byers, Robert S. Peabody Foundation for Archaeology (Kidder's Pecos Collection)

Edward B. Danson, Museum of Northern Arizona (Kiet Siel, Houck, Wupatki, Nalakihu, Winona, Ridge Ruin, and Pollock Ranch Collections)

Herbert W. Dick, Adams State College, Colorado, sponsored by Ft. Burgwin Research Center and the Pueblo of Picuris (Picuris Collection)

Charles C. Di Peso, Amerind Foundation, Inc. (Casas Grandes, Chihuahua, Mexico Collection)

Florence Ellis, University of New Mexico (Pojoaque Collection)

Rex E. Gerald and R. Roy Johnson, University of Texas at El Paso (Reeve Ruin Collection)

Raymond H. Thompson, Arizona State Museum (Point of Pines, Turkey Creek, and Gatlin Site Collections)

Alexander Wetmore, Neil M. Judd, and Richard L. Zusi, Smithsonian Institution and U. S. National Museum (Pueblo Bonito and Pueblo del Arroyo Collections)

My special thanks go to Mrs. Oran N. Robnett, Deming, New Mexico, for providing a macaw specimen and detailed information on its archaeological occurrence at the Freeman Ranch Site, near Cliff, New Mexico.

James Provinzano went to great lengths to obtain information pertaining to a macaw specimen from the Galaz Ranch Site, the only authenticated Military Macaw discovery in the entire Southwest.

The National Park Service loaned macaw specimens from its collections at the Southwest Archeological Center, Tuzigoot National Monument, and the Gran Quivira Project.

Joseph Forshaw, Canberra, New South Wales, Australia, provided important documentary sources; Kenton C. Lint, Curator, San Diego Zoological Gardens, and Jack Throp, Director, Honolulu Zoo, offered much-needed information on life histories of macaws; and Paul E. Violette undertook the translation from the French of important reference material. To all of these go my sincere thanks.

Four colleagues and old friends–Hildegarde Howard, Alexander Wetmore, Pierce Brodkorb, and Allan R. Phillips–whose opinions I value highly, agreed to undertake the critical reading of the manuscript (or sections thereof). I can find no way to thank them adequately for this onerous task other than to say I heeded their critical suggestions to the betterment of the paper. This does not, however, relieve me of the responsibility for its contents.

Throughout the study, Charmion McKusick served as my assistant. She prepared the illustrations and collated the material for the manuscript. I am deeply grateful to her for her skills and tireless efforts.

Other staff members of the Southwest Archeological Center also furnished indispensible special services and skills. Mrs. Lorrayne Langham typed the tabular measurement charts in their final form, and Mrs. Sue Bice, the final revised copies of the text. With unfailing good humor, Mrs. Annie Sanders, the Assistant Librarian, located much of the reference material needed in the study.

Finally, I am indebted to the University of Arizona for its sponsorship of the publication of this monograph.

L. L. H.

CONTENTS

ILLUSTRATIONS

TABLES

1. INTRODUCTION

The purpose of this paper is twofold, first to make possible the differentiation of the skeletal remains of the Military Macaw (*Ara militaris*) from those of the Scarlet Macaw (*Ara macao*), and second to review macaw remains recovered from archaeological sites. This paper is especially prepared for students of Southwestern archaeology because they are the ones who dig up macaw remains, who care for them in the field, who keep correlative field data for use in ethnological studies, and finally to whom may fall the task of determining economic values of macaws to the ancient inhabitants of now dead towns.

Historic references to these birds began about 1536 with the report by Cabeza de Vaca that Indians, living south of the present states of New Mexico and Arizona, "traded parrot's feathers for green stones far in the North" (Bandelier 1890: 61). Cabeza de Vaca did not describe the parrot he spoke of, so it may have been any one of 19 species of parrots found in Mexico, where there are large and small ones, long-tailed and short-tailed ones. The largest, which range in size (tip of bill to tip of tail) from 27 inches (about 685 mm) to 38 inches (about 960 mm), are called macaws (Blake 1953: 189-201).

Padre Luis Velarde was more specific than was Cabeza de Vaca. In 1716 he stated that "at San Xavier del Bac [Tucson, Arizona] and neighboring rancherías, there are many macaws, which the Pimas raise because of the beautiful feathers of red and of other colors, . . .which they strip from these birds in the spring, for their adornment" (Wyllys 1931: 129). These references testify to a desire for parrot feathers as a commodity for trade and also, specifically, to the raising of "red macaws" as a domestic source of macaw feathers for local use.

These uses of macaw feathers continued through the centuries, as shown in the Awatovi paintings of live macaws and of individual feathers (Smith 1952, Plate E). These illustrations and other data unquestionably prove possession of macaws and use of macaw feathers by Southwest Pueblo Indians from Pueblo II times to the 1960s, when the principal supplier of macaw feathers is the American aviculturist.

Macaw feathers and other "soft parts" are rarely found in archaeological situations outside of caves because of their perishable nature. So, evidence of the presence and use of macaws by prehistoric people, almost without exception, must come from a study of bone elements dug from the ground.

One of the primary purposes of the investigation of archaeological macaw remains is to determine uses that were made of live or dead macaws, whole or in part, by the inhabitants of a given place. Macaw remains are significant as products of the economy of the people, preserved through time to later serve in cultural interpretation. Further, they are proofs of contacts with distant peoples, possibly of different cultures. After death, the carcass of the macaw apparently no longer had economic value other than the possible use of feathers that might have been plucked from the dead bird before disposal. No artifacts made of macaw bones have yet been found in the Southwest, nor has any indication of the use of parrots for food been discovered.

2. MACAW IDENTIFICATION FACTORS

SIZE RANGE

It should be noted that size differences between the Mexican macaws and other Mexican parrots can be critical in distinguishing between them. At no period of development are the skeletal elements of the macaw small enough to fall within the size range of those of any other Mexican parrots (Fig. 1). See the paper by Olsen (1967) on the comparative osteology of macaws and thick-billed parrots, that appeared after this study was completed.

MORTALITY STATISTICS

In addition to feathers, which can be demonstrated to have been trade items in the prehistoric Southwest, the live birds themselves were items of commerce. The skeletons of these birds can produce useful data.

Observations on the age of the birds at death as determined by stages of osteological growth lead to inferences on the cultural role of the birds in economy, commerce, and ceremonial life.

To build an age-classification system based on skeletal characters, the following conditions must be met: (*1*) only macaws of a known age may be used, (*2*) only aviary bred macaws can provide a known age at death, (*3*) macaws are rare, and (*4*) cooperation is necessary from a breeder of macaws.

Different species of macaws of the genus *Ara,* having approximately the same size range, also seem to have approximately the same rate of bone growth. The age-classification system introduced and presented herein (Table 1) is thus a composite of the same age characters of the same skeletal elements of a total of 12 individuals, including the Blue-and-gold Macaw (*Ara ararauna*), the Military Macaw (*Ara militaris*), the Scarlet Macaw (*Ara macao*), and the Green-winged Macaw (*Ara chloroptera*). The Blue-and-gold Macaw and the Green-winged Macaw are not natives of Mexico.

In archaeological sites, specimens have been found from birds both younger and older than those listed in Table 1; these additional specimens constitute the

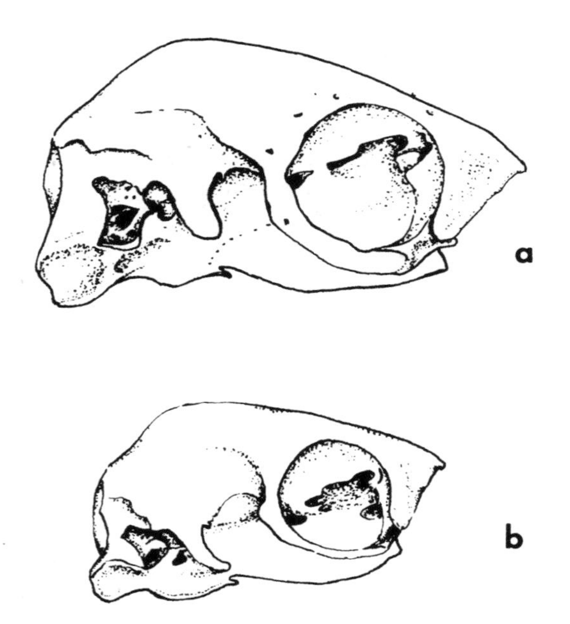

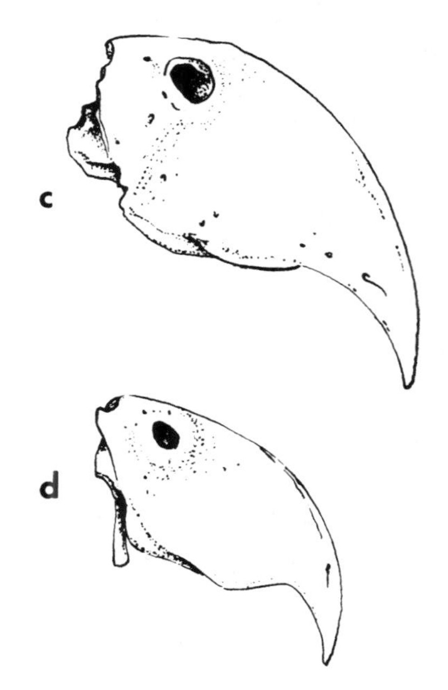

Fig. 1. Comparative sizes, in study specimen material, of the smallest macaw skull and premaxillary with those of the largest Mexican parrot of other genera. *a, Ara militaris* (H2440), cranium, lateral view; *b, Rhynchopsitta pachyrhyncha* (H2214), cranium, lateral view; *c, Ara militaris* (H2440), premaxilla, lateral view; *d, Rhynchopsitta pachyrhyncha* (H2214), premaxilla, lateral view. (Scale: X 1)

TABLE 1

Modern Macaw Specimens Used in Establishing Age Stage

Age stage	Specimen	Species	Age	Age Characteristics
Juvenile (7 wk – 4 mo)	H2451	*Ara ararauna*	7 wk	––
	H2449	*Ara ararauna*	8 wk	––
Immature (4 – 11 mo)	PB19338	*Ara macao*	Unknown	Between 4 & 11 mo
Newfledged (11 – 12 mo)	H2440	*Ara militaris*	11 – 12 mo	––
	H2444	*Ara chloroptera*	11 – 12 mo	––
	H2446	*Ara chloroptera*	11 – 12 mo	––
	Bi1351	*Ara macao*	Unknown	Same as for 11 – 12 mo
	H2159	*Ara macao*	Unknown	Same as for 11 – 12 mo
Adolescent (1 – 3 yr)	UA5386	*Ara militaris*	Unknown	Between 12 mo & 4 yr
	PB22561	*Ara macao*	Unknown	Between 12 mo & 4 yr
Breeding (4 - ? yr)	H2107	*Ara ararauna*	4+ yr	––
	H2099	*Ara macao*	Unknown	Same as for 4+ yr

Nestling and Aged groups. Study of this collection resulted in isolation of a series of diagnostic skeletal characters of age changes in macaws.

AGE CHARACTERS

Except in individuals less than eight weeks old, at which time the sternum is nearly all cartilage, the cranium, the sternum, the pelvis, and the tibiotarsus are the four bone elements that can be used reliably in determining the approximate age of a macaw at death. These defined characters, when grouped by age, also provide a classification of seven age stages: (1) Nestling, (2) Juvenile, (3) Immature, (4) Newfledged, (5) Adolescent, (6) Breeding, (7) Aged.

Nestling

Cranium – plates ossified in center only.
Sternum – cartilage only, unlikely in archaeological context.
Pelvis – vertebral column in four sections, sides not attached (Fig. 2*a*).
Tibiotarsus – neither proximal nor distal ends ossified, appearance stubby, swollen areas at ends and along fibular crest (Fig. 2 *b*).

Remarks: Age ranges from hatching to 7 weeks.

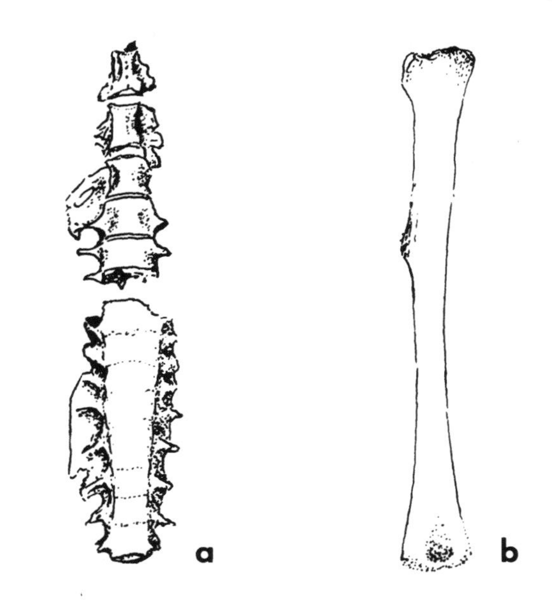

Fig. 2. Bone specimens from a Nestling, *Ara macao* (CG8,CG22), age about 6 weeks. *a,* synsacrum of pelvis; *b,* right tibiotarsus, anterior view. (Scale: X 1)

By about 3 weeks the tarsometatarsus has ossified sufficiently to indicate the angle and direction of the outer proximal foramen, thus providing a character for species determination at this early age (Fig. 3).

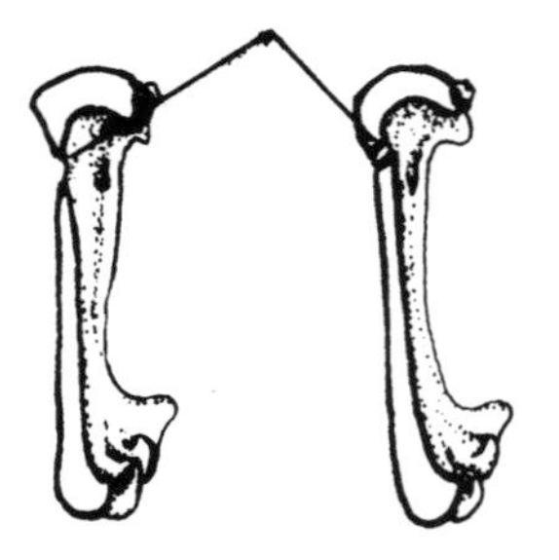

Fig. 3. Proximal external foramen from a Nestling (CG8,CG22). Left *Ara militaris,* right *Ara macao,* longitudinal cross sections. (Scale: X 1)

Juvenile

Cranium – at 7 weeks, thin plates usually transparent and separate, only bones around *foramen magnum* fused (Fig. 4*a, b*); at 8 weeks, posterior third fused to bones around *foramen magnum* (Fig. 5*a, b*); at 4 months, cranium one unit, less suborbital arch (Fig. 6*a, b*).

Sternum – at 7 weeks, still all cartilage except bone tissue forming at sternal rib attachments (Fig. 4*c*); at 8 weeks, anterior third ossified (Fig. 5*c*); at 4 months, sternum completely ossified (Fig. 6*e*).

Pelvis – at 7 weeks, completely fused, surface texture less porous, smoother (Fig. 4*e*).

Tibiotarsus – at 7 weeks, distal head ossified (Fig. 4*d*); at 4 months, proximal head ossified, tendinal bridge not yet formed (Fig. 6*f*).

Remarks: Age ranges from 7 weeks to 4 months. In formation of the cranial cavity, two fenestrae are left in the ventral surface of the cranium (Fig. 6*d*); these remain open until adulthood. The surface appearance of the pelvis is less porous than formerly, becoming more solid and shiny until mature. By 4 months, swelling at ends and along fibular crest of the tibiotarsus has nearly subsided (Fig. 6*f*).

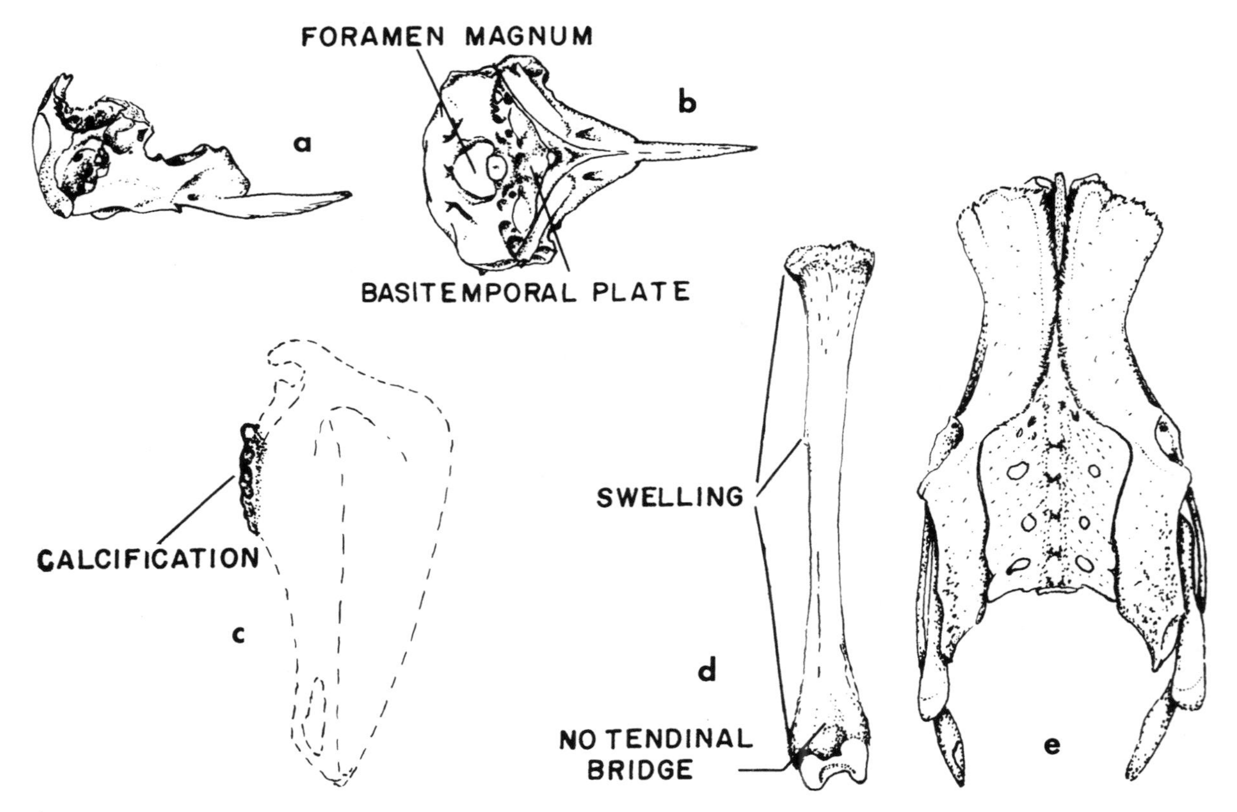

Fig. 4. Bone specimens from a Juvenile, 7 weeks old, *Ara ararauna* (H2451). *a, b,* cranium, lateral and ventral view; *c,* sternum, lateral view; *d,* right tibiotarsus, anterior view; *e,* pelvis, dorsal view. (Scale: X 1)

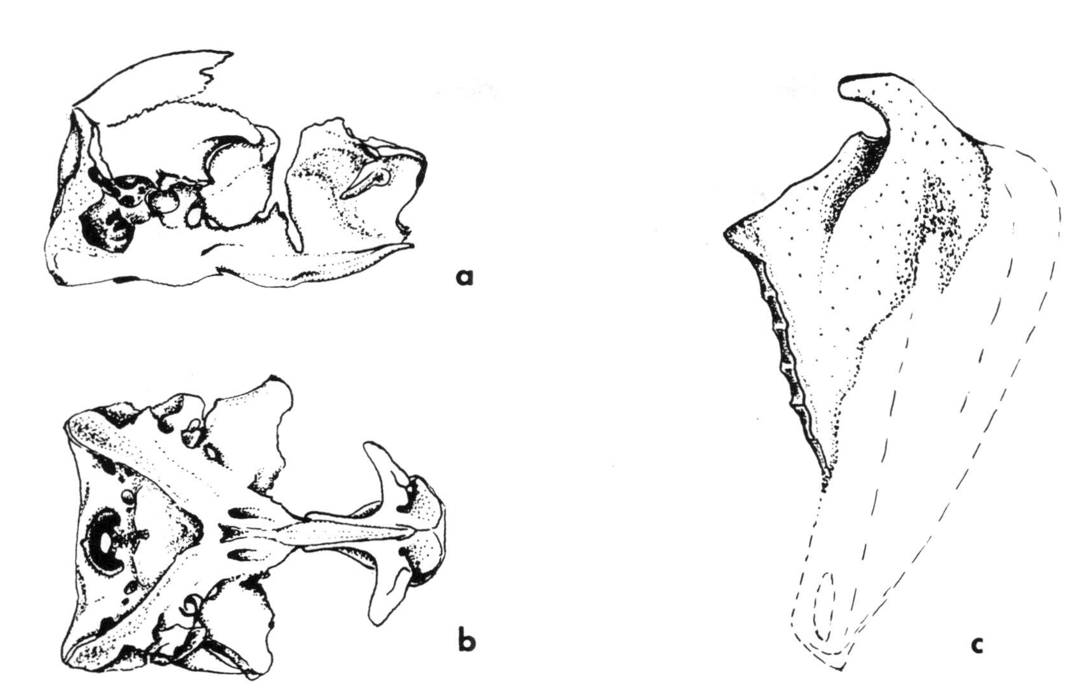

Fig. 5. Bone specimens from a Juvenile, 8 weeks old, *Ara ararauna* (H2449). *a,* cranium, lateral view; *b,* cranium, ventral view; *c,* sternum, lateral view. (Scale: × 1)

Immature

Cranium – by 4 months opaque, suborbital arches lengthened but not complete (Fig. 6*a, b*); fenestrae still open (Fig. 6*d*).
Sternum – completely ossified (Fig. 6*e*).
Pelvis – completely ossified (Fig. 6*g*).
Tibiotarsus – tendinal bridge still incomplete (Fig. 6*f*).

Remarks: Age ranges from 4 to 11 months.

Newfledged

Cranium – by 11 months, suborbital arches complete (Fig. 7*a*), fenestrae still open (Fig. 7*b*).
Sternum – completely ossified (Fig. 7*c*).
Pelvis – completely ossified (Fig. 7*e*).
Tibiotarsus – tendinal bridges still incomplete (Fig. 7*d*).

Remarks: Age ranges from 11 to 12 months.

Adolescent

Cranium – by 13 months suborbital arches complete (Fig. 8*a*); fenestrae growing smaller along posterior and medial margins (Fig. 8*b*, 9*a*).
Sternum – completely ossified.
Pelvis – completely ossified.
Tibiotarsus – tendinal bridge complete (Fig. 8*c*).

Remarks: Age ranges from 13 months through third year. From Adolescence (1-3 years) on, the raised edge of the basitemporal plate becomes accentuated.

Breeding Age

Cranium – 4 to ? years, suborbital arches complete (Figs. 7*a*, 8*a*), fenestrae half closed (Fig. 9*b*).
Sternum – completely ossified.
Pelvis – completely ossified.
Tibiotarsus – tendinal bridge complete, and broad (Fig. 11).

Remarks: Age ranges from 4 to ? years. The term "breeding" to denote an age category was selected because macaws, at least the large-size aviary birds, are usually five years old before chicks capable of survival are produced. Confirmation has been made by Kenton C. Lint, Curator, San Diego Zoological Gardens, who wrote me in 1966: "We have had some Macaws lay eggs at the age of two years. Most of our birds have not had viable chicks until five years of age. Other pairs did not nest until ten years of age."

Substantial agreement was provided by Jack L. Throp, Director, Honolulu Zoo: "Macaws probably don't breed before they are three years old."

Both of these opinions are well substantiated in conversation with Bernard Roer, Roer Bird Farm, Phoenix, Arizona, that at four years macaws may lay eggs, but are not able to produce viable young until they are five years of age. Moreover, Roer said that some of his individuals have refused to accept a mate until associated for as much as five years. "Age 4 to ? years" now seems a fair range for "Breeding Age."

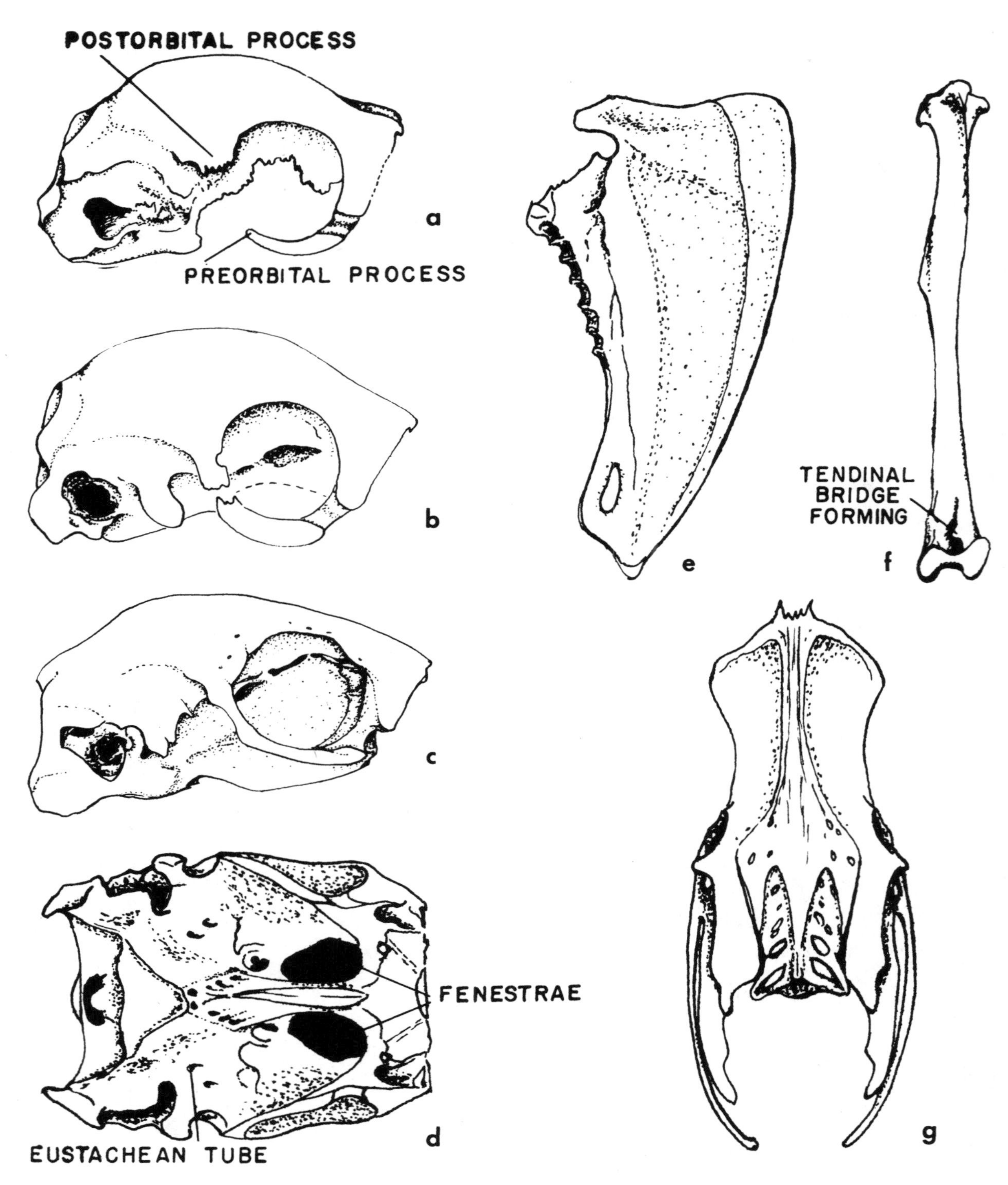

Fig. 6. Bone specimens from an Immature, 4-11 months, *Ara macao* (CG298, CG246, H2107, PB19338). *a,* cranium, lateral view, early Immature; *b,* cranium, lateral view, late Immature; *c & d*, cranium, lateral and ventral view of Newfledged, showing fenestrae, 11-12 months; *e,* sternum, lateral view; *f,* right tibiotarsus, anterior view; *g,* pelvis, dorsal view. (Scale: X 1)

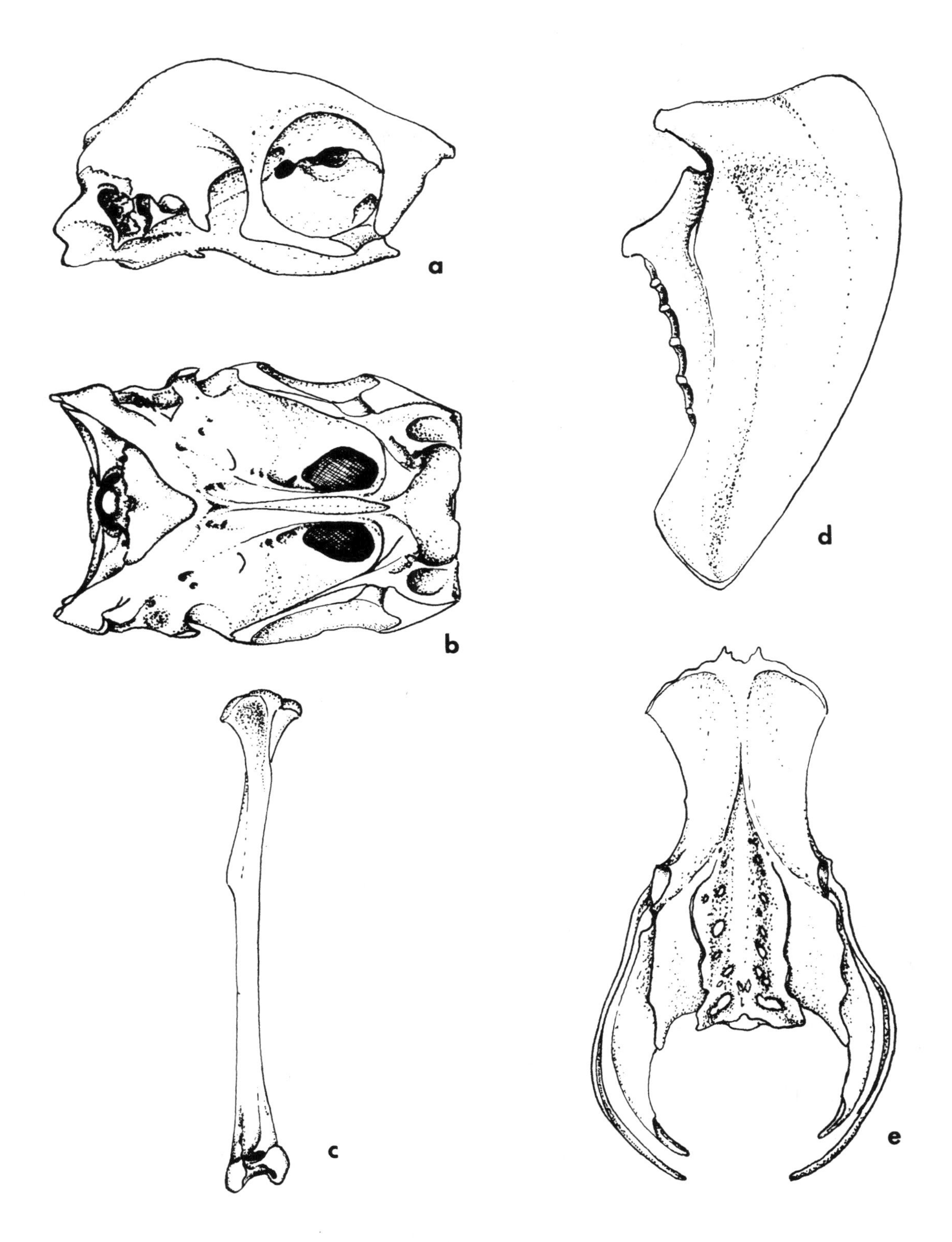

Fig. 7. Bone specimens from a Newfledged, 11 to 12 months, *Ara macao* (H2159). *a*, cranium, lateral view; *b*, cranium, ventral view; *c*, right tibiotarsus, anterior view; *d*, sternum, lateral view; *e*, pelvis, dorsal view. (Scale: × 1)

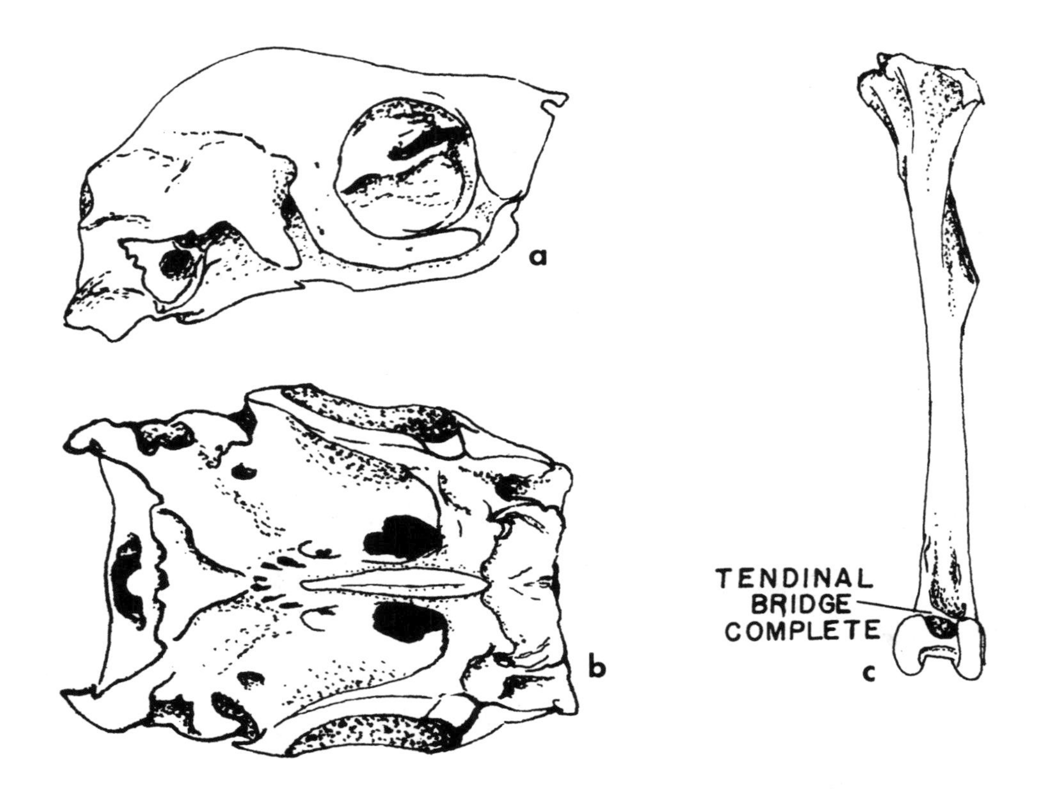

Fig. 8. Bone specimens from an Adolescent, 1-3 years, *Ara macao* (PB22561). *a,* cranium,. lateral view; *b,* cranium, ventral view; *c,* left tibiotarsus, anterior view. (Scale: X 1)

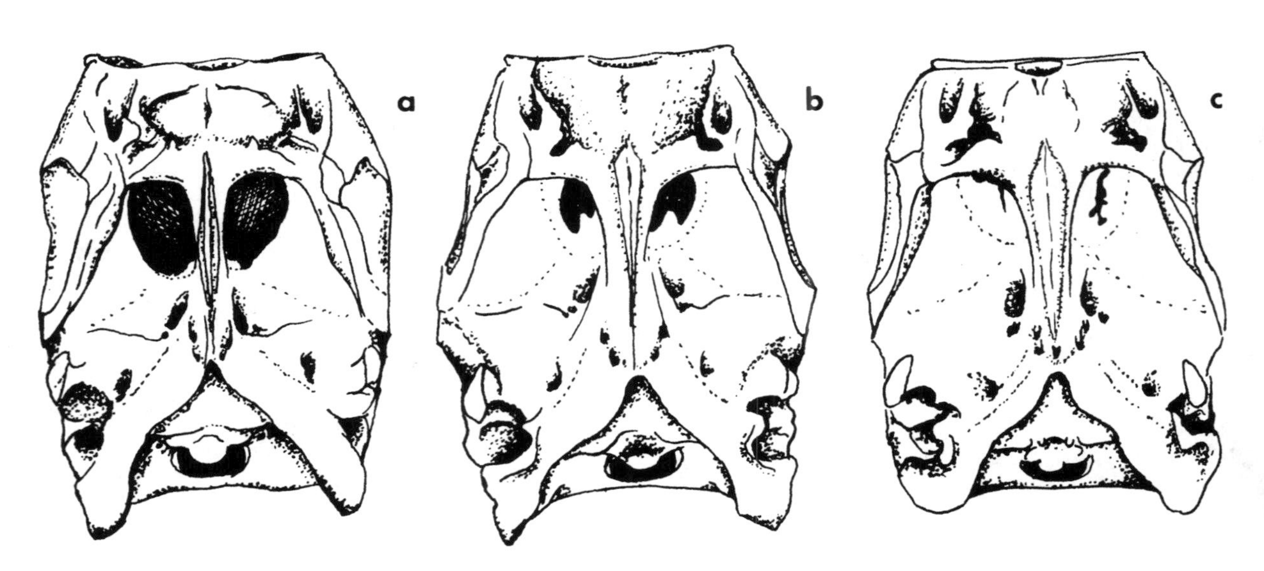

Fig. 9. Comparison of crania of Adolescent (1-3 years), Breeding Age (4+ years) and Aged. *Ara macao* (CG140, CG216.1, CG229). *a,* cranium, ventral view, Adolescent; *b,* cranium, ventral view, Breeding Age; *c,* cranium, ventral view, Aged. (Scale: X 1)

Aged

Cranium – extremely old, suborbital arches complete, fenestrae almost completely closed (Fig. 9*c*).
Sternum – completely ossified.
Pelvis – completely ossified.
Tibiotarsus – tendinal bridge complete and broad (Fig. 11).

Remarks: It should not be difficult to recognize bones of very old macaws because of progressive changes such as those illustrated in Figure 10. It would be exceptional to locate an extremely old modern macaw with a known age at the time of its death; however, it is known that macaw "individuals have lived upwards of 50 years in captivity" (Austin 1961: 146). Of special interest is a Military Macaw that was given by Neil M. Judd to a Zuni Indian in 1924. This macaw, denuded of feathers at intervals, died in 1946 (Judd 1954: 263), which indicates that a macaw can survive for at least 22 years as a producer of feathers under conditions essentially comparable to those to which archaeological macaws were subjected. The age of macaws at death thus is important in archaeological interpretation.

Where age characters for the genus *Ara* are identified and defined, distinctions between species of the genus can be made. An example may be used in which an age character was not recognized as such

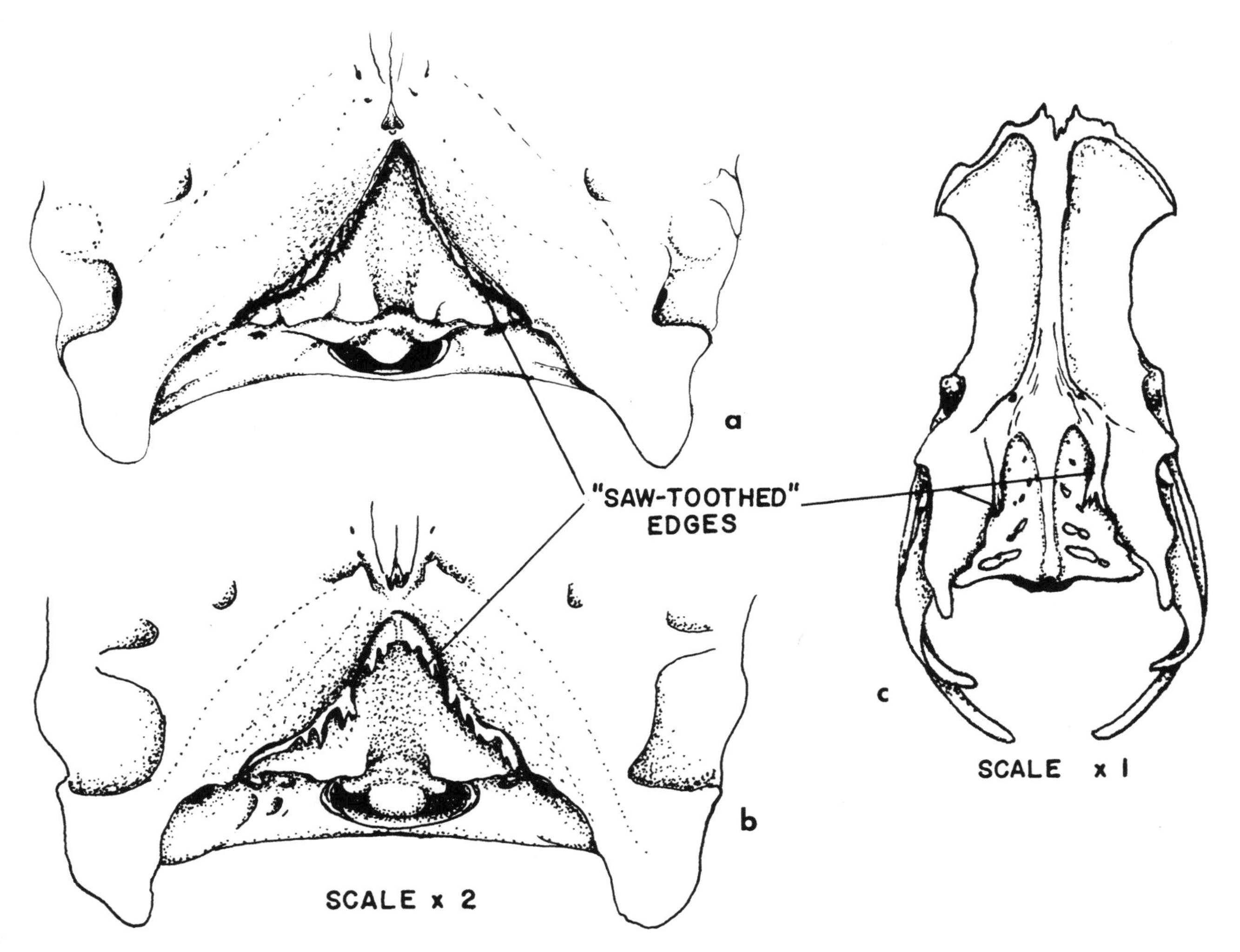

Fig. 10. Manifestations of age, the development of "saw-toothed" edges on the basitemporal plate of the cranium and of the interior iliac margins of the pelvis. *a, Ara macao* (PB22561), Adolescent (Scale: X 2); *b, Ara militaris* (CG76.1), Adolescent (Scale: X 2); *c, Ara macao* (PB22561), Adolescent (Scale: X 1).

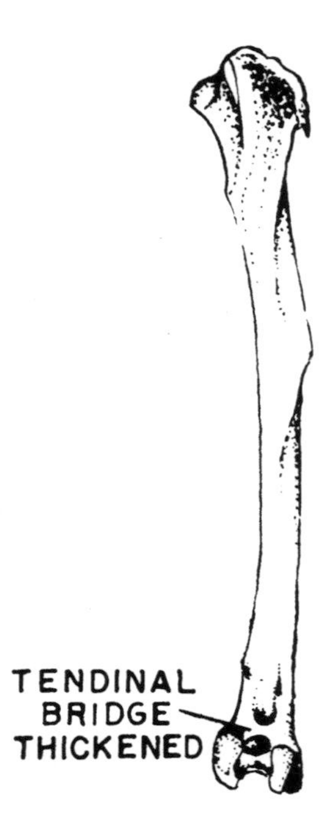

Fig. 11. Broadened tendinal bridge. *Ara macao* (H2099) left tibiotarsus, anterior view. (Scale: X 1)

and was mistaken for a species character. Referring to the Blue-and-gold Macaw (*Ara ararauna*), Thompson (1899: 9, 27) has commented that the differences of orbital ring are well known, "though imperfectly investigated," a fact he later demonstrated when he said that "the skull of *A. ararauna* is remarkable in having the orbital ring incomplete, though the long curved prefrontal processes approach very close to the postfrontal; it is complete in the other . . . ."

According to my findings his specimen was an Immature, less than a year, probably 9 months old; hence his example was not a valid one for his study. If this macaw had been 3 months older the orbital rings would have been complete as in the skulls of the other species of the genus he studied.

RANGE OF THE MEXICAN MACAWS

Macaws are not widespread over Mexico, although the Military Macaw is generally and sparsely distributed over the arid and semiarid regions in pine and oak forests up to 8,000 feet. Its nearest occurrence to the United States is in Sonora, more than 175 airline miles south of the border at Soyopa, at 28° 45′ N. latitude, noted by Wright (October 14, 1931) as "by far the most northerly point at which the Military Macaw has been detected" (Van Rossem 1945: 101).

In contrast, the Scarlet Macaw is restricted to humid lowland forests from southern Tamaulipas on the east coast and Oaxaxa on the west coast south beyond Mexico (Blake 1953: 192).

DESCRIPTIONS OF SPECIES

For the purpose of identifying an individual macaw there are two formal descriptions–one based on external characters and the other on skeletal characters. External characters are those used in describing the form and beauty of a bird, as the Military Macaw, and they provide the bases for the written description universally accepted as "the description of a bird" in bird-guides and in other publications as well. This description tells us what the bird looks like in life. The two Mexican Macaws are described as follows.

Military Macaw (*Ara militaris*)

External Description: ". . .27-30 [inches]. Mainly green and yellowish olive-green, the forehead and lores bright red; remiges (above), middle and greater wing-coverts purplish blue like tips of tail; rump and tail-coverts cerulean blue; tail very long and pointed, the basal half deep red above, the underside uniform yellowish olive-green like under side of wings; bare sides of head pinkish; iris yellow" (Blake 1953: 191).

Scarlet Macaw, formerly "Red-Blue-and-Yellow Macaw" (*Ara macao*)

External Description: ". . . 34-35 [inches]. Extensively scarlet or vermilion, the wing-coverts rich chrome-yellow, the remiges deep purplish blue; lower back, rump and tail-coverts azure-blue; tail very long, pointed and mainly red; bare sides of head pinkish; iris yellow" (Blake 1953: 191-2).

SPECIES IDENTIFICATION

External descriptive characters are normally relied upon in identifying a bird species, or in defining differences between distinctive characters of more than one species. Plumage colors and patterns of the Military and Scarlet macaws are so conspicuously different that each species is readily identified from these characteristics. Few studies of the osteology of two species of this genus have been made, and there is no literature on the subject.

In the case of archaeological or paleontological specimens, skeletal elements must be relied upon for species identification. Character for character an unknown bone must be compared to the same bone element of a known species. It is advisable to have a sufficiently large series of skeletons of each macaw species involved, in order to determine the regularity of occurrence of a given character. As such series are rarely accessible, the diagnostic key for identification of bones of *Ara militaris* and *Ara macao* is based on seven skeletons as follows:

Bi1351 – Military Macaw (*Ara militaris mexicana*), female. Wild bird shot February 18, 1938, at Tenacatita Bay, Jalisco, Mexico. Newfledged; age at death 11-12 months. Loaned by the Los Angeles County Museum of Natural History.

H2440 – Military Macaw (*Ara militaris*), female. Nestling hand-raised in captivity. Died December 13, 1965. Newfledged; thought to be 11 months old at death. Ova 17 x 4 mm. Molting. Hargrave Collection.

UA5386 – Military Macaw (*Ara militaris mexicana*), female. Wild bird shot February 21, 1964, 30 miles northeast of Mazatlan cut-off on Mexico Highway 40. Sinaloa, Mexico. Collected by Richard S. Crossin, Original No. 794. Ova 14 x 3 mm. Adolescent; age at death 1-3 years. University of Arizona Collection.

H2099 – Scarlet Macaw (*Ara macao*). Date of death? Breeding Age; age at death 4 to ? years. Molting. Reported to probably be trade stock from South America to Tepic, Nayarit, Mexico. Donated by Lewis Yeager. Hargrave Collection.

H2159 – Scarlet Macaw (*Ara macao*), female. May 1963 died in captivity. Newfledged; age at death 11-12 months. Acquired from Roer Bird Farm. Aviary bred. Ova 16 x 6.5 mm. Molting, Hargrave Collection.

PB19338 – Scarlet Macaw (*Ara macao*). Immature; age at death 10-11 months. Collection of Pierce Brodkorb.

PB22561 – Scarlet Macaw (*Ara macao*). Adolescent; age at death 1-3 years. Collection of Pierce Brodkorb.

Before characters referable to each of these species could be separated from the mass of characters involved, it was essential that age characters, as previously defined, be eliminated from consideration. Of the remaining characters some are referable to higher taxonomic levels (taxons), others to sex differences, and still others to individual variation. Segregation of sex characters was attempted, but small sample-size and extent of overlap make the data unusable at this time. In this study each diagnostic character for one species is a configuration of bone describably different from its homologue in the other species (Table 2, Figs. 12, 13).

RESIDUAL CHARACTERS
(Variations from the norm)

The remaining characters, of unknown significance, must be referred to the genus *Ara*. Some of these characters may originate with sex, as do spurs in turkeys, which in turn create related characters. Other characters may be the effects of ecology or of a changed personal environment; character alterations may result from mild but persistent human pressure, as restricted activities, and from severe pressure resulting in broken bones. Malnutrition can produce changes in the bone, thus creating abnormal configurations. The origin of configurations should be understood for proper evaluation of a character in relation to a specialized study. All of these possibilities should be considered in determining age and species. Figure 14 shows four humeri that differ conspicuously one from the other; but the diagnostic species character, the external-bicipital-curve to shaft, immediately proves that all four specimens represented are the Scarlet Macaw (*Ara macao*). In considering the question of domestication, Delacour (1964: 154) writes:

> We call domestic birds those which not only have been brought to live and breed regularly under man's control and artificial conditions, but also have changed in proportions, size, life habits and other characteristics so that they have become more useful to man in supplying him with meat, eggs and down.

These traits vary in populations; therefore, native raised birds should be compared with aviary bred stock.

COMPARATIVE BONE MEASUREMENTS

In order to compare birds from archaeological sites with modern stocks, tables of measurements have been compiled, and ranges in size have been calculated for each measurement (Tables 2-8). Scarlet Macaws have been segregated from those individuals too fragmentary for species identification. Bones so pathological as to give an unreliable measurement have been excluded.

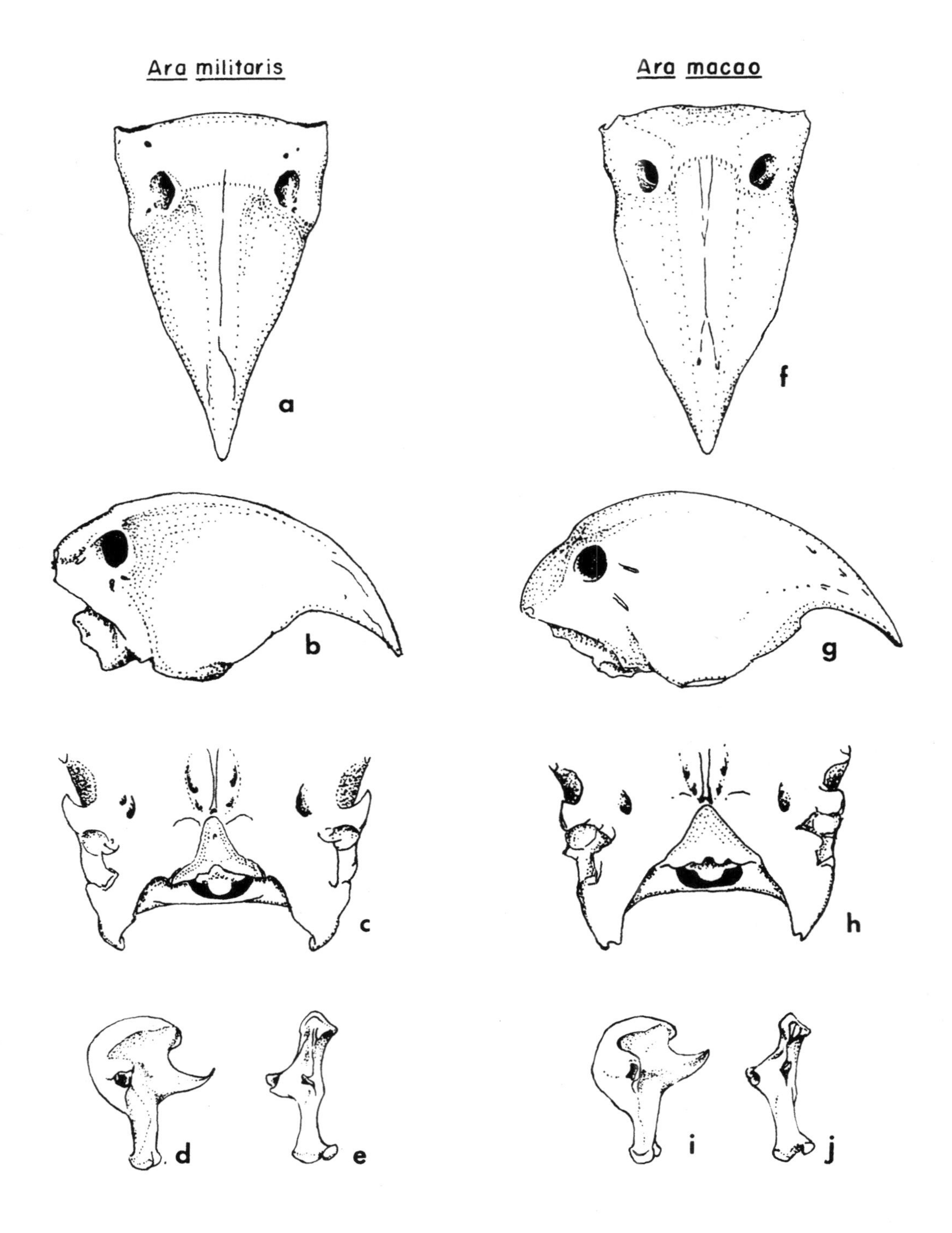

Fig. 12. Comparison of bones of (*a* - *e*) *Ara militaris* (UA5386 and Bil351) and bones of (*f* - *j*) *Ara macao* (H2099 and H2159). *a,* premaxilla, frontal view; *b,* premaxilla, lateral view; *c,* basitemporal plate, ventral view; *d,* left quadrate, lateral view; *e,* left quadrate, anterior view; *f,* premaxilla, frontal view; *g,* premaxilla, lateral view; *h,* basitemporal plate, ventral view; *i,* left quadrate, lateral view; *j,* left quadrate, anterior view. (Scale: X 1)

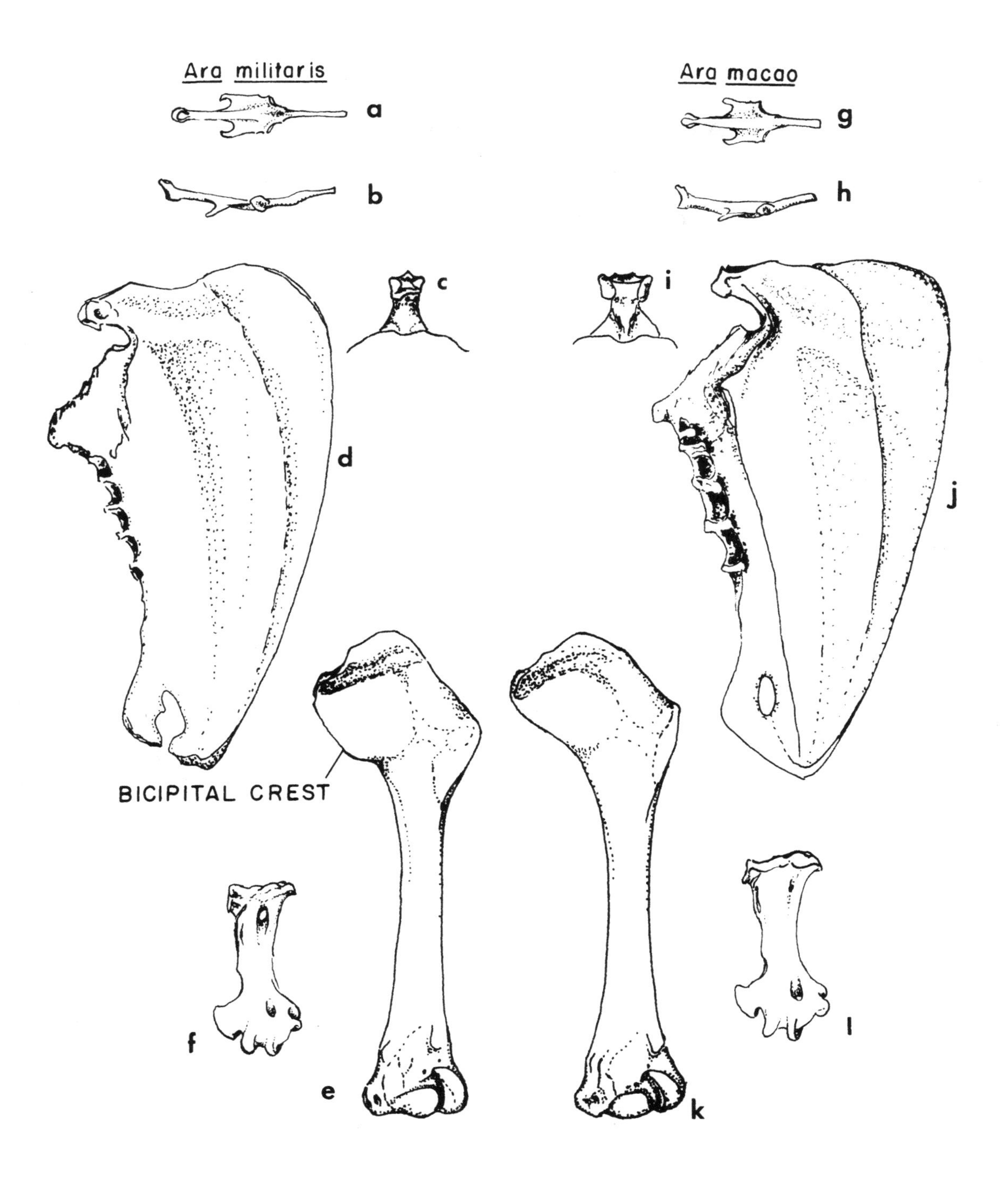

Fig. 13. Comparison of bones of (*a* - *f*) *Ara militaris* (H2440 and UA5386) and bones of (*g* - *l*) *Ara macao* (CG230 and PB22561). *a,* basihyal-basibranchial, dorsal view; *b,* basihyal-basibranchial, lateral view; *c,* manubrium, dorsal view; *d,* sternum, lateral view; *e,* left humerus, palmar view; *f,* left tarsometatarsus, anterior view; *g,* basihyal-basibranchial, dorsal view; *h,* basihyal-basibranchial, lateral view; *i,* manubrium, dorsal view; *j,* sternum, lateral view; *k,* left humerus, palmar view; *l,* left tarsometatarsus, anterior view. (Scale: X 1)

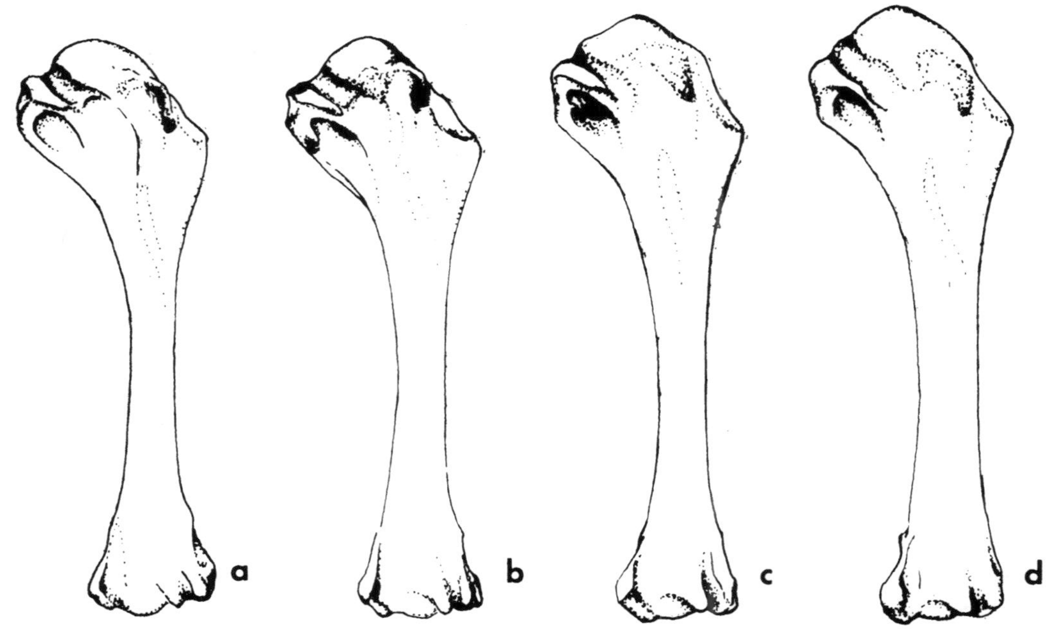

Fig. 14. Variation within a species, *Ara macao,* anconal view. *a,* (H2159), modern aviary bird; *b,* (H2099), modern aviary bird; *c,* (C5), archaeological; *d,* (V39) archaeological. (Scale: X 1)

TABLE 2

Diagnostic Characters for Modern *Ara militaris* and *Ara macao*

Element	Ara militaris	Ara macao
Premaxilla	Line of culmen narrow, acute-angled, flat-sided (Fig. 12 *a, b*)	Line of culmen broad, well-rounded slightly inflated (Fig. 12 *f, g*)
Cranium	Basitemporal plate "bell-shaped" (Fig. 12 *c*)	Basitemporal plate triangular (Fig. 12 *h*)
Quadrate	Orbital and otic processes longer and more slender (Fig. 12 *d, e*)	Orbital and otic processes short, stocky (Fig. 12 *i, j*)
Basihyal-basibranchial	Basihyal slender, basibranchial long and tapering (Fig. 13 *a, b*)	Basihyal stocky, basibranchial short (Fig. 13 *g, h*)
Sternum	Tip of ventral manubrial spine (dorsal view) irregular, convex, narrow (Fig. 13 *c*)	Tip of ventral manubrial spine (dorsal view) regular, broad and bifurate (Fig. 13 *i*)
	Posteriorly angled, anterior carinal margin sharply indented (Fig. 13 *d*)	Anterior carinal margin straighter (Fig. 13 *j*)
Humerus	Junction of bicipital crest and shaft forms definite angle (Fig. 13 *e*)	Bicipital crest curves regularly into the shaft (Fig. 13 *k*)
Tarsometatarsus	Outer proximal foramen large, penetrates shaft at right angle to axis of shaft (Fig. 13 *f*)	Outer proximal foramen penetrates shaft (posterior view), angling distally (Fig. 13 *l*)

TABLE 3

Description of Bone Measurements Used in Tables 4 Through 8

Number	Bone	Description
1	Premaxilla	maximum width hinge line
2	Cranium	maximum axial length of crown
3	Cranium	minimum interorbital width of crown
4	Cranium	minimum width suborbital bridge
5	Cranium	maximum width eye socket
6	Cranium	maximum depth eye socket
7	Cranium	maximum depth at anterior point of basitemporal plate
8	Scapula	maximum length
9	Scapula	maximum width coracoidal head
10	Coracoid	maximum axial length
11	Coracoid	maximum width sternal attachment
12	Furcula	maximum length
13	Furcula	maximum width
14	Pelvis	maximum width across antitrochanters
15	Humerus	maximum length
16	Humerus	maximum width proximal head
17	Humerus	maximum width distal head
18	Ulna	maximum length
19	Ulna	maximum width, proximal head, palmar view across condyles
20	Ulna	maximum width distal end internal view
21	Ulna	maximum width distal end anconal view
22	Radius	maximum length
23	Radius	maximum width distal end
24	Carpometacarpus	maximum length
25	Carpometacarpus	maximum width proximal end internal view
26	Carpometacarpus	maximum width distal end at symphysis
27	Femur	maximum axial length
28	Femur	maximum width distal end
29	Femur	maximum width proximal end posterior view
30	Femur	maximum depth distal end posterior view
31	Tibiotarsus	maximum length
32	Tibiotarsus	maximum width proximal end posterior view
33	Tibiotarsus	maximum width distal end anterior view
34	Tarsometatarsus	maximum axial length
35	Tarsometatarsus	maximum width distal end, anterior view, axial postion
36	Tarsometatarsus	minimum width shaft, anterior view, through metatarsal facet

TABLE 4

Bone Measurements (in millimeters) of Modern Macaw Specimens Used in Establishing Age Stages (See Table 3 for description of measurements)

Species and Specimen Number	Pre-maxilla	Cranium						Scapula		Coracoid		Furcula		Pelvis	Humerus		
	1	2	3	4	5	6	7	8	9	10	11	12	13	14	15	16	17
Ara militaris																	
LACM B:1351	32.6	61.6	46.8	5.5	20.4	18.9	33.1	57.7	13.9	54.7	13.8	37.2	27.6	38.7	83.2	21.8	16.8
H2440	31.1	58.2	41.5	2.9	20.0	19.6	30.6	53.2	13.1	51.6	13.5	33.7	23.7	36.5	75.9	20.5	16.0
UA5386	32.1	59.0	39.2	3.1	22.3	20.4	31.1	54.2	13.2	53.6	15.0	32.4	23.5	36.2	79.3	22.0	16.4
Ara macao																	
H2159	31.7	59.6	42.8	3.1	21.8	20.5	30.7	55.7	13.6	56.1	14.1	34.1	24.1	34.4	79.9	20.3	16.1
H2099	31.5	55.8	42.1	2.7	19.2	19.1	32.0	54.6	13.1	55.4	15.1	35.8	22.1	36.6	79.9	20.6	16.1

TABLE 5

Bone Measurements (in millimeters) of Scarlet Macaws Found at Pueblo Bonito by Pepper* (See Table 3 for description of measurements)

Site and Specimen Number	Pre-maxilla	Cranium						Scapula		Coracoid		Furcula		Pelvis	Humerus		
	1	2	3	4	5	6	7	8	9	10	11	12	13	14	15	16	17
Pueblo Bonito																	
(C)5226	31.3	58.3	43.0	3.1	18.9	18.6	31.4	56.4	13.6	51.9		33.7	23.3	39.0	81.3	22.3	17.1
(C)5227	32.2	59.5	46.1	4.0	20.4	19.6	34.6	56.7	13.3	53.9	15.8	35.9	22.8	40.0	84.1	23.3	17.8
(C)5228	31.8	59.4	42.6	3.5	20.0	19.4	30.2	57.2	13.7	55.6	16.1	33.3		38.9	85.4	23.6	17.4
(C)5229	29.6	57.1	39.9	4.6	18.8	19.1	30.3		13.9	54.2	14.6	35.2	22.0	39.3	84.4	22.5	17.5
(C)5231	30.7	58.7	42.0	3.6	19.8	19.9	30.2	56.0	12.7	53.2	14.3	36.9	24.9	37.7	84.3	23.1	17.8
(C)5232	31.0	58.0	43.9	2.8	20.0	19.6	30.4	56.4	14.1	51.2	14.8			40.0	82.4	21.6	17.1
(C)5233	31.6	57.5	41.9	3.9	19.2	18.5	32.7	58.1	13.7	52.0	15.8				83.2	21.5	16.5
(C)5234	30.1	59.8	42.6	4.1	19.7	19.9	31.8	54.3	12.3	53.1	15.4				83.9	22.1	17.4
(C)5235	31.9	56.6	42.2	3.5	19.9	19.3	32.3	54.5	13.0	53.5	13.3				80.5	20.9	16.3
(C)5236		61.3	43.7	3.6	21.9	20.2	33.1	54.9	12.9	53.6	15.3				82.3	22.1	16.9
(C)5237		58.1	40.2		20.5		30.3			55.0	15.3				85.0	21.3	17.4
(C)5238	30.3	58.4	43.5	3.4	19.8	18.9	32.4								81.0	21.7	17.1
(C)5239															80.8	21.0	16.2
Range	29.6 to 32.2	56.6 to 61.3	39.9 to 46.1	2.8 to 4.6	18.8 to 21.9	18.5 to 20.2	30.2 to 34.6	54.3 to 58.1	12.3 to 14.1	51.2 to 55.6	13.3 to 16.1	33.3 to 36.9	22.0 to 24.9	37.7 to 40.0	80.5 to 85.4	20.9 to 23.6	16.2 to 17.8

* Bones of cranium from known individual birds; other bones mixed.

Ulna				Radius		Carpometacarpus			Femur				Tibiotarsus			Tarsometatarsus		
18	19	20	21	22	23	24	25	26	27	28	29	30	31	32	33	34	35	36
102.1	12.3	10.6	12.1	94.0	8.4	62.9	15.7	9.4	61.2	13.0	13.3	9.6	84.7	14.5	10.8	30.8	15.3	6.2
	11.6	10.3	11.3	86.5	8.3	57.9	15.1	9.2	56.8	12.4	12.8	8.8	80.2	14.1	10.6	29.0	14.5	5.6
98.9	11.6	10.5	11.6	91.0	8.4	61.2	15.2	9.2	58.5	12.2	12.9	9.4	81.2	14.3	10.6	30.5	14.5	5.1
102.7	12.1	10.2	11.3	95.8	8.3	61.1	15.6	9.4	60.0	12.6	13.9	9.8	85.2	14.3	10.8	30.9	14.6	5.3
102.3	11.8	10.0	11.4	95.0	8.1	62.5	15.1	9.1	59.1	11.9	13.1	8.7	83.0	14.6	10.0	29.7	14.3	5.7

Ulna				Radius		Carpometacarpus			Femur				Tibiotarsus			Tarsometatarsus		
18	19	20	21	22	23	24	25	26	27	28	29	30	31	32	33	34	35	36
104.6	12.0	10.2	11.6	95.7	7.7	64.5	15.8	9.5	61.7	13.3	14.4	10.2	84.7	14.1	11.6	29.9	15.5	7.1
105.9	12.0	10.7	11.6	94.2	8.0	64.4	15.8	9.8	57.6	12.2	12.8	9.7	84.1	14.3	10.7	31.9	15.1	6.1
106.7	12.4	10.3	11.3	93.4	8.7	63.9	16.5	9.7	60.3	12.8	13.4	9.9	85.1	14.5	11.4	30.8	14.9	5.8
106.2	12.6	10.6	11.3	97.8	8.4	59.8	16.6	9.8	60.8	13.7	14.4	10.2	84.2	13.6	11.3	31.4	14.1	6.0
98.6	12.6	10.8	11.8	91.8	7.6	63.7	16.4	10.2	60.8	12.9	14.0	9.8	83.4	12.8	10.5	30.2	15.0	6.5
108.4	12.3	10.8	11.7	91.3	8.7	62.2	15.8	9.3	59.7	13.0	14.0	9.8	79.7	13.1	11.4	30.0	14.9	6.3
104.9	12.1	10.9	10.6	97.8	8.6	65.6	16.1	9.6	57.7	12.5	13.4	9.3	84.8	14.8	11.2	31.7	14.9	6.7
102.7	11.7	10.1	11.5	97.2	8.8	64.1	15.5	9.7	61.6	12.5	13.2	9.7	81.9	13.9	10.9	29.8	14.5	6.4
100.2	11.8	9.7	10.9	96.1	8.0	61.2	14.1	9.6	57.1	12.0	12.8	9.7	84.5	14.6	11.4	29.8	15.2	5.8
99.9	11.8	9.7	10.6	96.4	7.9	64.0	15.7	9.8	59.3	13.5	13.9	10.6	81.0	13.6	10.3	32.9	15.1	6.3
104.3	13.0	11.1	12.0	95.2	7.3	60.9	15.7	9.4	60.6	13.2	13.6	10.3	80.7	14.1	11.7			
102.4	12.0	10.8	11.4			62.4	15.2	9.9	59.4	12.2	13.2	9.6		14.4	11.8			
105.5	12.6	11.2	12.0						60.5	13.9	14.4	11.0						
98.6	11.7	9.7	10.6	91.3	7.3	59.8	14.1	9.3	57.1	12.0	12.8	9.3	79.7	12.8	10.3	29.8	14.1	5.8
to	to	to	to	to	to	to	to	to	to	to	to	to	to	to	to	to	to	to
108.4	13.0	11.2	12.0	97.8	8.8	65.6	16.6	10.2	61.7	13.9	14.4	11.0	85.1	14.8	11.8	32.9	15.5	7.1

TABLE 6

Bone Measurements (in millimeters) of Other Scarlet Macaw Archaeological Specimens (See Table 3 for description of measurements)

Site and Specimen Number	Premaxilla	Cranium						Scapula		Coracoid		Furcula		Pelvis	Humerus		
	1	2	3	4	5	6	7	8	9	10	11	12	13	14	15	16	17
Pueblo Bonito																	
H/6452	31.2	69.9	45.2	3.8	19.5	18.1	30.5		13.7	51.5					81.5	28.0	17.2
H/6708	31.1	59.7	42.2	3.7	20.5	20.0	30.0	56.0	13.6						79.4	27.8	16.9
H/6709		56.2	44.7	5.0	18.6	18.7	31.0	54.3	13.8	55.0					83.6	29.3	16.5
343578	27.7	57.0													82.7	22.9	16.8
343576	29.1	58.5	44.5	2.7	18.9	19.2	30.6			53.6	13.9				82.9	21.7	17.2
343571	32.5	61.9	44.9	3.7	20.5	20.6	33.0	59.3	14.8	56.4	14.9				86.7	23.8	17.9
343572	32.8	59.3	45.3	3.5	20.0	20.1	31.6	52.8	13.8	53.9	14.6				83.2	22.3	16.3
343573A		60.6	46.7	3.9	19.2	18.9	33.1	53.2	13.2	52.4	13.0				80.7	21.2	16.0
343573B																	
343581							30.6		14.2	52.2	14.5				84.4	22.4	16.9
343579		58.6	42.5	4.2	19.9	19.8				54.3	14.8				79.7	22.9	17.2
343574	34.6	60.8	45.9				31.2			51.2					80.7	22.9	16.7
343575	29.2	59.1	42.1	3.4	21.8	19.9	30.6								77.4	21.8	16.4
343583	29.5	59.4	43.6	3.5	19.1	19.7	32.9										
343580								57.0	14.2	52.4	14.8				79.8	21.4	17.0
343580								56.8	13.8						86.5	23.6	17.9
343580																23.1	
Pueblo del Arroyo																	
C5	27.8	57.3	42.5	2.9	19.9	18.6		55.0	13.3	52.9	15.0	34.1	25.5	34.7	82.3	21.3	16.2
344360	31.2	59.2	45.9	3.7	19.9	19.5	34.7		13.2		14.7			39.1	82.2	22.4	16.9
344359	31.0	57.5	41.2	3.9	19.6	19.0				54.5	14.5				85.2		16.6
Kin Kletso																	
C163	30.3	59.1	42.8	4.4	19.1	18.6	32.4										
Houck K																	
A0.804																	
Freeman Ranch																	
D2221		60.4		4.1	19.7	20.5	34.4								87.4	23.7	18.3
Wupatki, NA 405																	
A0.544		56.4	38.9				30.1										
A0.548		61.0	45.4	4.0	19.1	18.7	31.7										
A0.533															79.9	22.0	16.0
A0.538	30.6		45.9						13.3								16.8
A0.535									14.2						88.5	22.6	18.3
A0.537	34.2														86.3	22.4	18.0
A0.534	29.3																
A0.528/ W99	32.0						32.8			52.9		36.1	27.8			25.0	

Table 6 (Continued ⟶)

Ulna				Radius		Carpometacarpus			Femur				Tibiotarsus			Tarsometatarsus		
18	19	20	21	22	23	24	25	26	27	28	29	30	31	32	33	34	35	36
102.3	12.0	11.4	9.5	94.4	8.4	62.0	15.8	9.5	57.0	13.1	13.1		83.0	14.0	10.5	29.6	15.1	6.6
99.2	12.1	11.5	9.4	91.3	8.2	60.7	15.5	9.7	59.8	13.5	13.4		81.3	14.5	11.1	30.5	14.8	7.7
106.7	10.8	11.5	9.4	98.0	8.2	65.6	15.6	9.3	58.1	12.5	13.0		82.7	14.0	10.7	31.7	14.9	5.8
104.8	12.4	10.5	11.4	95.4	7.8	63.3	15.6	9.9	60.9	13.2	13.5	9.9	84.3	14.1	11.4	31.2	16.2	6.8
107.2	13.1	10.8	12.0	98.5	8.7	62.9	16.0	10.1	62.2	13.4	13.8	10.2	86.7	15.1	11.3	31.9	15.2	6.1
104.0	12.6	10.4	11.6	95.3	8.2	63.6	15.9	10.1	60.8	12.8	13.9	10.1						
100.9	11.7	10.3	11.2	93.1	7.6	61.9	14.8	9.0										
104.8	12.1	10.0	10.8	96.0	8.2	63.0	15.6	9.3	61.3	12.2	13.2	9.7	85.6	13.0	11.2	31.5	14.4	5.5
100.8	12.3	10.7	10.8	93.2	8.0	62.2	16.0	9.9	59.5	13.7	14.7	10.4	82.2	14.5	12.0	31.6	16.3	6.0
102.0	12.3	10.7	11.5	92.9	8.6	61.9	15.3	9.2	60.1	13.1	13.8	10.5	85.9	14.8	11.5	30.8	15.0	6.8
94.2	11.7	10.1	11.0						55.9	12.0	13.1	9.1				28.6		6.0
101.6	12.2	10.7	11.7	98.7	8.5	63.1	15.7	9.7	60.1	12.8	13.2	9.2	85.2	13.9	11.1	31.1	15.2	6.1
106.9	13.1	10.8	11.7	97.2	8.5	62.3	15.8	9.7	64.7	14.0	14.7	10.7	88.6	15.0	11.9			
106.0	12.6	11.4	12.1	93.2	8.6													
	11.8						14.8		59.1	12.2	13.0	9.1	80.6	13.1	10.0	30.4	13.3	5.7
103.6	12.5	10.3	11.5	95.1	8.5	62.6	15.6	10.0	59.5	12.7	13.9	9.9	83.1	13.9	11.0	30.0	14.8	6.7
106.0	12.3	10.6	11.7	97.8	8.3	66.1	15.7	9.4	59.9	12.5	13.4	9.4	84.2	14.2	10.6	32.0	14.2	6.0
109.7	13.1	10.0	12.2	100.4	8.5				62.6	13.2	14.0	9.8						
99.6	11.9	10.1	11.2															
	12.3	10.2	10.8		7.7	63.6	15.0	9.6	58.0	12.5	13.7	9.5	81.8	14.5	10.8			
109.8		12.0	12.5	102.5	9.7	66.2	16.8	10.2	64.9	14.0	14.6	9.6	99.9	15.8	12.3			6.9
108.8	12.8		12.6			66.1	16.3	10.3					88.1	14.7	10.4			
									58.5	12.7	13.2	8.9						

Table 6 (Continued)

Site and Specimen Number	Premaxilla	Cranium						Scapula		Coracoid		Furcula		Pelvis	Humerus		
	1	2	3	4	5	6	7	8	9	10	11	12	13	14	15	16	17
Wupatki, NA405 (Continued)																	
W84	33.2	61.2	45.1	4.4	20.0	19.7	33.3										
W50		58.3	42.0	3.8			31.0		13.3					39.9	82.0	22.5	17.8
W105	33.5	59.8	42.6		20.8		31.7								83.3	23.1	16.6
A0.540	30.1		43.2	3.5	18.1	18.4	31.9	55.7	14.4	54.7	14.8				83.1	21.8	18.0
A0.542		55.3							13.2	52.8	15.3				79.7		16.4
A0.325															77.2	21.0	15.7
A0.80															82.1	22.2	17.1
A0.549		58.7	43.6	3.6	20.1	18.9	32.2										
A0.547		58.8	44.7	3.1	19.7	18.6	31.4										
A0.536									14.4						83.8	21.7	16.8
A0.81																21.4	
A0.539	31.2	62.1			21.0		32.0	56.4	13.2	55.5	14.9				84.8	22.1	17.6
A0.764																	
A0.773																	
Ridge Ruin, NA 1785																	
A0.467															82.6		17.6
A0.468																	
A0.497																	
Pollock Site, NA 4317																	
A0.710															84.3	22.8	17.2
Tuzigoot																	
V958	31.0	58.2	44.7	3.4	20.5	19.2	32.3	57.5	13.4	54.6	14.0			40.4	85.4	22.6	17.1
V39															73.9	21.9	17.2
V40															84.4	22.6	17.6
Montezuma Castle																	
J514															84.4	22.2	17.2
Jackson Homestead																	
J121		61.2	44.7	4.7	19.7	19.0	32.1										
Kinishba																	
D813	30.5	56.3	40.3	3.7	19.3	19.2	28.7	53.8	13.2	50.8	14.7				76.6	21.2	15.9
D814	30.1	60.0	40.5	3.7	20.1	19.4	34.6		13.1	55.7	15.7			39.8	87.7	22.6	17.5
D815	32.7		44.2				30.1		14.0	54.1	15.0	34.0	31.9		83.3	22.2	17.0
D830		57.2	41.0	3.1	21.2	19.2	31.9										
Point of Pines, Ariz. W:10:50																	
D2179															78.8	20.8	16.3
D2180			39.0				30.6										
D2181	30.7	56.7	43.2	3.1	19.7	18.6	31.5		12.8	53.4	16.3			39.6	78.9	21.8	16.7
D2182																	
D2184				3.4	19.5	18.6								39.0	83.0	20.8	16.9

Table 6 (Continued →)

Ulna				Radius		Carpometacarpus			Femur				Tibiotarsus			Tarsometatarsus		
18	19	20	21	22	23	24	25	26	27	28	29	30	31	32	33	34	35	36
100.7	12.6	10.9	11.4		8.7	63.7	16.9	9.4	58.8	13.1	14.0	9.7	81.5	14.8	12.3	30.9	14.9	6.8
102.8	12.3	10.9	11.9	93.5	8.0	62.3	15.8	9.6	61.2	13.3	13.7	9.9		14.6	10.8	32.4	15.8	6.2
101.8	12.9	10.5	11.8	93.2	8.7	62.2	16.2	9.9	60.9	13.1	14.0	8.9	84.2	15.1	11.5	32.3	15.7	6.4
97.8	11.9	10.1	11.1	88.7	7.1	60.4	15.1	9.2			13.1							
104.8	12.3	10.8	12.3			62.5	15.4	10.2										
104.2	12.3	10.6	11.4	95.9	7.9	63.8	15.3	10.0	61.3	12.9	13.3	8.9	96.7		11.0			
																31.3	15.1	5.8
																30.9	15.2	5.9
	12.6	10.7	12.1	95.4	8.3	63.1	16.3		59.7	12.5	13.1	9.8				29.7	15.1	6.3
																31.8	15.5	6.2
94.9	12.4	10.5	11.2	97.1	8.1	65.0	15.9	9.5										
105.0	12.7	10.5	11.9	97.2	8.6				58.8	14.2	13.4		83.4	15.0	11.1			
104.5	12.5	10.3	9.6			63.0	16.1	9.8					85.8	13.8	10.8			
97.9	11.5	9.7	9.0	90.3	7.6	59.8	14.4	8.8	57.3	12.6	13.0	9.4	79.6	13.1	10.1	30.5	14.4	5.4
105.5	12.3	10.4	10.6	97.8	8.0	63.3	15.8	9.8	63.9	13.1	14.0	9.8	86.8	13.9				
101.1	12.1	11.1	11.4	92.3	7.8	62.6	15.9	9.6	59.6	12.4	13.0	10.3	82.5	14.0	10.9	30.8	15.1	5.6
99.2	11.8	10.0	11.4	90.2	7.7	60.9	15.8	9.9	56.5	12.3	13.0	8.8	77.7	13.4	11.5	29.4	14.5	5.5
104.3	12.5			95.5		61.9	15.3	9.5								30.3	13.4	5.5

Table 6 (Continued)

Site and Specimen Number	Pre-maxilla	Cranium						Scapula		Coracoid		Furcula		Pelvis	Humerus		
	1	2	3	4	5	6	7	8	9	10	11	12	13	14	15	16	17
Point of Pines, Ariz. W:10:50 (Continued)																	
D2186			39.8	3.0	19.5	17.7	31.4								83.7	20.4	17.1
D2187		57.7	42.6	2.8	19.7	18.8	31.7	53.7	13.5	53.8	15.8						
D2189																	
D2190				4.3	18.7	19.7	31.8	55.0	13.4	51.1	13.9				78.4	21.1	15.8
D2194	33.6	59.9	46.2	3.4	20.0	18.4	33.0	55.7	15.9								
D2196				4.4				59.3	14.4	55.8	17.1				86.0	23.3	17.6
D2198	30.1														80.7	21.9	16.8
D2200																	
D2203		57.5	42.4	3.8	19.1	19.2											
D2205			57.3		19.3	19.6									81.6	22.0	16.9
Turkey Creek, Ariz. W:10:78																	
D2208		59.7	46.5		19.4		32.6										17.0
D2209				4.1	19.9		32.6										
D2210		61.3	46.2	3.4	20.2	20.5	33.3								83.7	22.9	17.8
D2212		60.0	48.1	4.2	19.9	18.6	32.6										
D2213		61.0	45.8	4.9	18.8	18.0									86.9	22.4	17.7
D2214															88.0	22.0	17.4
D2217	34.1	60.9	47.7	3.7	20.9	18.9	31.4	57.3	13.9	56.1	15.0				86.4	22.9	17.7
Galaz Site																	
D2269																	
D2270	33.6	57.4	42.6		19.6		31.9										
D2271	32.8	60.2	44.6												79.8		17.7
Gatlin Site, Arizona 2:2:1																	
D2005																	
Pecos																	
291284A	33.1	60.2	42.4	3.1	19.8	20.2	31.3	56.2	13.6	56.6	15.9	35.2	24.2		86.9	23.3	17.7
291284B	31.6	58.5	41.7	3.8	19.2	18.4				54.8	15.4				79.7	21.7	16.9
Gran Quivira																	
Q4	31.1	56.4	43.8	3.1	19.9	18.5	31.7	56.4	13.7	52.1	16.3			36.7	81.9	22.8	17.3
García Site																	
D766	28.1																
Range	27.8 to 34.2	55.3 to 69.9	38.9 to 48.1	2.8 to 5.0	18.6 to 21.2	17.7 to 20.5	28.7 to 34.4	53.7 to 59.3	13.2 to 15.9	51.5 to 56.6	14.0 to 17.1	34.0 to 36.1	24.2 to 31.9	34.7 to 40.4	73.9 to 88.5	20.4 to 29.3	15.9 to 18.3

Ulna				Radius		Carpometacarpus			Femur				Tibiotarsus			Tarsometatarsus		
18	19	20	21	22	23	24	25	26	27	28	29	30	31	32	33	34	35	36
104.9	12.0	10.4	11.4	97.9	7.9				60.7	12.5	12.0	10.1	85.0	13.8	11.0			
						62.7	14.8	9.6	60.5	13.3	13.5	9.6	84.3	13.9	11.6	30.4	15.0	6.1
													79.4	14.0	10.7	30.9	16.9	7.2
97.8	11.4	10.2	11.0	90.4	7.7	60.9	16.0	9.8	56.5	11.8	12.5	8.5	79.5	13.4	10.7	30.0	13.3	5.6
107.2	12.7	10.7	11.8	98.8	8.4	64.5	16.2	9.9	63.4	13.0	14.3	10.3	89.4	14.6	11.5	31.8	15.0	6.4
106.2	12.6	10.9	11.9	97.4	8.6	65.0	15.9	10.3	61.3	13.7	14.6	10.0	86.0	14.7	11.5	31.4	16.3	6.9
													80.3	14.0	11.2			
															10.7	31.8		5.6
															11.2			
104.6	12.3	10.1	11.3						57.3	12.8	13.2	9.4						
					7.7				56.9	12.2	12.3	8.9	84.2	14.8	10.8			
107.0	13.2	10.7	12.2	98.9	8.4													
106.4	12.3	10.8	11.6															
				99.4	8.8													
100.2	12.6	10.3	11.3			61.2	14.6	9.7				9.8						
						58.5		9.4								31.1	15.9	6.2
101.6	12.3	10.3	11.5						57.5	12.6	13.4	10.4	80.8	15.1	10.8			
	12.7	11.2	11.9	97.6	8.4	64.6	15.9	10.4	61.2	13.2	14.3	9.9		14.3				
99.6	12.5	10.6	12.0	91.0	7.6	61.6	15.6	9.6	56.6	12.0	12.6	9.5	76.7	13.9	10.5	30.4	14.2	5.8
102.6	12.7	10.5	11.7		8.1	61.5	16.3	10.8	59.4	13.6	13.8	10.3	82.9	14.6	12.3	29.7		6.7
94.2 to 109.8	10.8 to 13.2	9.7 to 12.0	9.0 to 12.6	88.7 to 102.5	7.1 to 9.7	59.8 to 66.2	14.4 to 16.9	8.8 to 10.3	56.5 to 64.9	11.8 to 14.6	12.0 to 14.6	8.5 to 10.3	79.4 to 99.9	13.1 to 15.8	10.0 to 12.3	29.4 to 32.4	13.3 to 16.9	5.4 to 7.2

TABLE 7

Measurements (in millimeters) of Three Bones of the Military Macaw From the Galaz Site (See Table 3 for description of measurements)

Specimen Number	Cranium 2	3	4	5	6	7	Ulna 18	19	20	21	Tarsometatarsus 34	35	36
D2268						31.7	96.5	11.7	10.1	11.3	28.3	14.5	5.7

TABLE 8

Bone Measurements (in millimeters) of Macaw Archaeological Specimens Not Identified by Species (See Table 3 for description of measurements)

Site and Specimen Number	Pre-maxilla 1	Cranium 2	3	4	5	6	7	Scapula 8	9	Coracoid 10	11	Furcula 12	13	Pelvis 14	Humerus 15	16	17
Kiet Siel																	
A0.437	30.7	57.9	42.9	2.6	21.2	19.4	30.3	54.6	13.1	52.5	14.6			36.6	8.0	21.7	15.8
A0.52																	
Pueblo Bonito, Judd																	
USNM 343577	32.5		43.1				32.4							41.8			
Nalakihu, NA 358																	
A0.326																	
Wupatki, NA 405																	
A0.755																	
A0.160																	
A0.323																	
A0.529																	
A0.752																	
A0.541				3.5				57.8	14.8	57.1	15.8	37.6	26.2				
A0.87																	
A0.79																	
A0.531																	
A0.324																	
A0.285																	
A0.555																	
A0.556																	
A0.183																	
A0.322																	

Table 8 (Continued →)

Ulna				Radius		Carpometacarpus			Femur				Tibiotarsus			Tarsometatarsus		
18	19	20	21	22	23	24	25	26	27	28	29	30	31	32	33	34	35	36
98.2	11.6	9.9	8.3	89.6	7.8	58.6	14.5	8.8	56.7	12.3	12.9	9.1	80.6	14.0	10.2	29.4	14.8	5.8
													87.4	14.4				
						59.3	14.6	9.4										
						63.2	15.5	10.0										
						64.9	15.9	9.9	62.8	13.1		10.7	86.9	14.6	11.7			
		11.1	12.0														15.6	
															11.1			
															10.6			
104.2	12.1	10.4	11.4															
		10.7	11.5															

Table 8 (Continued)

Site and Specimen Number	Pre-maxilla	Cranium						Scapula		Coracoid		Furcula		Pelvis	Humerus		
	1	2	3	4	5	6	7	8	9	10	11	12	13	14	15	16	17
Wupatki, NA 405 (Continued)																	
A0.50																	
A0.766																	
A0.753																	
A0.757																	
Ridge Ruin, NA 1785																	
A0.494																	
Winona, NA 3644																	
A0.426																	
Point of Pines, Ariz. W:10:50																	
D2183																	
D2185																	
D2188				3.3													
D2191								54.7	13.4								
D2192																	
D2193																	
D2195																	
D2197																	
D2199																	17.0
D2201																	
D2202																	
D2204		59.3			20.9												
Point of Pines, Ariz. W:10:78																	
D2206																	
D2207																	
D2211																	16.0
D2215																	
D2216																	
Reeve Ruin																	
D2220																	
Picuris Pueblo																	
PP424																	

Ulna				Radius		Carpometacarpus			Femur				Tibiotarsus			Tarsometatarsus		
18	**19**	**20**	**21**	**22**	**23**	**24**	**25**	**26**	**27**	**28**	**29**	**30**	**31**	**32**	**33**	**34**	**35**	**36**
		10.9	11.4															
															10.7			
									58.6	12.9	14.1							
													81.4	13.5	11.2			
										13.9	14.3	9.5						
100.1		10.3	11.6				15.7						82.7	12.9	11.8			
99.3	12.2	10.4	11.8	92.3	8.1	61.5	15.9	10.6										
103.5	12.1	9.6	10.9	95.4	8.4								81.8	14.4	12.3			
98.7		9.8																
													83.6	14.4	10.9			
													81.6	13.6	11.3			
													82.0					
						59.1	15.4	9.1										
													85.1	14.9	11.0			
													79.7	12.9	11.1			
		10.6	11.5		8.1													
		10.1	11.2															

3. CULTURE AREAS PROVIDING MACAW REMAINS

Macaw remains of one kind or another have been found associated with cultural remains of each of the basic cultures of the Southwest (Figs. 15, 16). Since distinctions in ceramic characters can be readily made in order to correlate macaws with cultures, an abstract of diagnostic ceramic characters is listed herein where appropriate, usually near the beginning of a cultural discussion.

ANASAZI CULTURE AREA

The Anasazi were mainly concentrated in the drainage of the San Juan River in the general Four Corners area. Their ceramic cultural indicators are (*1*) applied decoration–paint (black), (*2*) surface color–normally shades of gray or white, (*3*) surface finish utility vessels – frequently corrugated, (*4*) firing atmosphere – reducing, and (*5*) method of thinning vessel walls – not satisfactorily established.

Chaco Canyon, New Mexico

In the middle section of the Chaco River, San Juan County, Northwestern New Mexico, in the drainage of the San Juan River, are several large, ruined communal dwellings up to five stories in height, within the boundary of Chaco Canyon National Monument. In three of these long-abandoned pueblos were found remains of macaws.

Pueblo Bonito

Apparently the first archaeological parrot bone from a prehistoric human habitation of the southwestern United States was recovered from Pueblo Bonito by George H. Pepper, Field Director of the Hyde expeditions, sometime during the late 1890s (Pepper 1920: 1, 195). The discovery of parrot skeletons and indications that the birds had been kept in cages (Pepper 1920: 194) was in keeping with knowledge of that century that parrots were kept by modern Pueblo Indians of New Mexico, as implied by Espejo in 1583 who "found a parrot in a cage" while among the Keres Indians (Bolton 1916: 181). The recovery of parrot skeletons from Pueblo Bonito may thus have been expected.

In Room 38 (Pepper 1920: 193-4) Pepper found his first macaws, two "that had died and been buried" and a larger group that had been killed in a roof fall. These bones have been restudied, with the result that all 13 individuals are now identified as *Ara macao* (Scarlet Macaw). Nine were Newfledged (11-12 months), two were Adolescent (1-3 years), one was Aged, and one too incomplete to age. The Aged bird had survived a broken leg and cranial deformity during its lifetime. These birds display indications of premature old age, a metabolic deficit contributed to by lack of sunshine and proper food. Eight of the 13 had roughened ulnae.

Bones Recovered by Pepper

The bones of the macaws from Room 38 of Pueblo Bonito were mixed with those of other birds and mammals; therefore a brief summary of individuals is presented here.

(c) 5226 – Scarlet Macaw (*Ara macao*), Newfledged (11-12 months)
Provenience: Room 38
Elements: Nearly complete skeleton

(c) 5227 – Scarlet Macaw (*Ara macao*), Newfledged (11-12 months)
Provenience: Room 38
Elements: Nearly complete skeleton

(c) 5228 – Scarlet Macaw (*Ara macao*), Adolescent (1-3 years)
Provenience: Room 38
Elements: Nearly complete skeleton

(c) 5229 – Scarlet Macaw (*Ara macao*), Newfledged (11-12 months)
Provenience: Room 38
Elements: Nearly complete skeleton

(c) 5231 – Scarlet Macaw (*Ara macao*), Adolescent (1-3 years)
Provenience: Room 38
Elements: Nearly complete skeleton

(c) 5232 – Scarlet Macaw (*Ara macao*), Newfledged (11-12 months)
Provenience: Room 38
Elements: Nearly complete skeleton

(c) 5233 – Scarlet Macaw (*Ara macao*), Newfledged (11-12 months)
Provenience: Room 38
Elements: Nearly complete skeleton

(c) 5234 – Scarlet Macaw (*Ara macao*), Newfledged (11-12 months)
Provenience: Room 38
Elements: Nearly complete skeleton

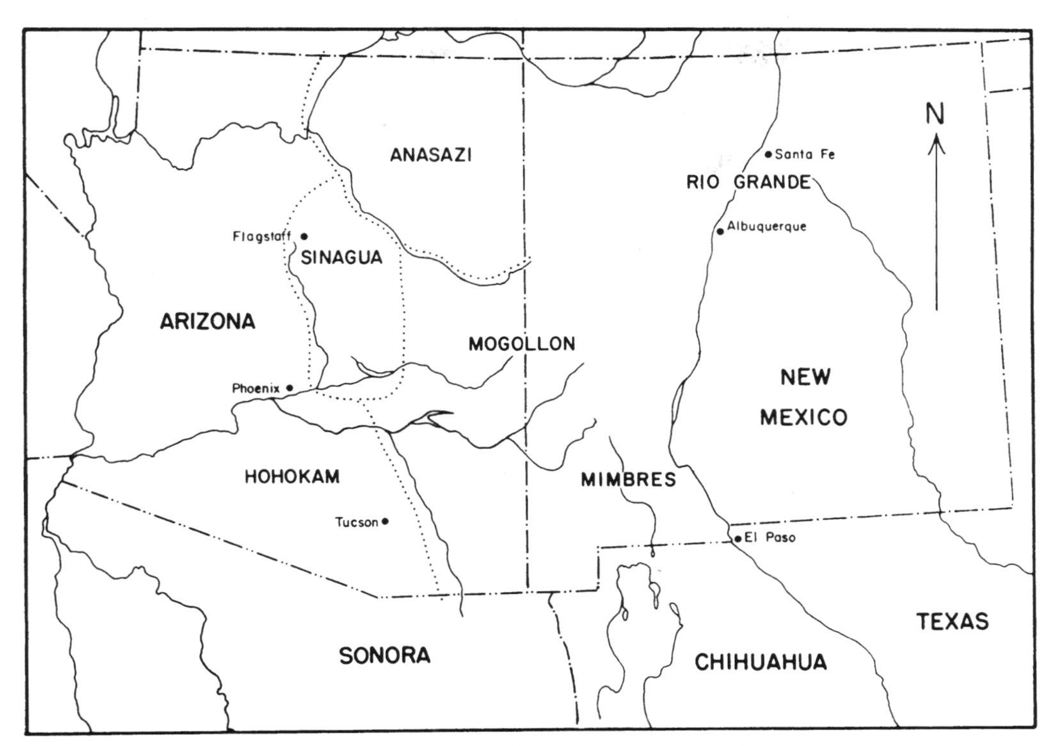

Fig. 15. Culture areas of the prehistoric Southwest.

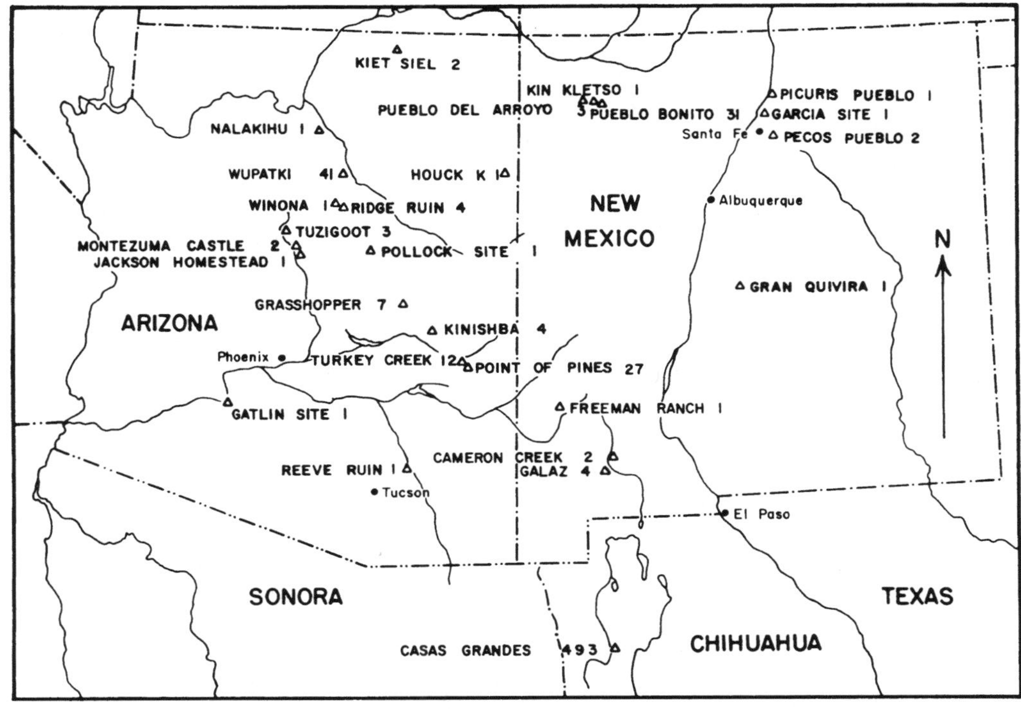

Fig. 16. Distribution of macaw archaeological remains. Numbers indicate number of birds found at site.

(c) 5235 – Scarlet Macaw (*Ara macao*), Newfledged (11-12 months)
Provenience: Room 38
Elements: Nearly complete skeleton

(c) 5236 – Scarlet Macaw (*Ara macao*), Newfledged (11-12 months)
Provenience: Room 38
Elements: Nearly complete skeleton

(c) 5237 – Scarlet Macaw (*Ara macao*), Aged
Provenience: Room 38
Elements: Nearly complete skeleton

(c) 5238 – Scarlet Macaw (*Ara macao*), Newfledged (11-12 months)
Provenience: Room 38
Elements: Nearly complete skeleton

(c) 5239 – Scarlet Macaw (*Ara macao*)
Provenience: Room 38
Elements: Humeri, ulnae, femora

Comments. The measurements taken from the bones recovered from Room 38 suggest a macaw population having short, deep crania and rather long wings, a physical configuration which may indicate a source of supply from an area widely separated from that of most of the other macaws studied. A computer might help establish this variation as a significant difference. It is with this hope that the measurements for this group are presented separately (Table 5).

Over 20 years after Pepper's excavation, in 1921, Neil M. Judd, as Director of the National Geographic Society Expedition, began extensive archaeological investigations at Pueblo Bonito; these continued seasonally until the fall of 1927 (Judd 1954: vii). Judd (1954: 263) found remains of macaws of "no less than 10 articulated skeletons and a number of miscellaneous bones."

The identification by Alexander Wetmore (Judd 1954: xi) of the Scarlet Macaw (*Ara macao*), a native of humid jungles hundreds of miles to the south, failed to stir up much curiosity or interest about where it came from and how it got there. Years have dulled unwritten factual details of occurrence, so proveniences are known for only a few of the following 16 individuals obtained for restudy.

Bones Recovered by Judd

USNM343577 – (No. 9) – Macaw (*Ara sp.*), Immature (4-11 months).
Provenience: Room 255.
Elements: Mandible, eight tracheal rings, and pelvis.
Remarks: Aged on basis of mandible.

H/6452 – Scarlet Macaw (*Ara macao*), Adolescent (1-3 years old). Species identification on premaxilla, basitemporal plate, basihyal-basibranchial, quadrate, manubrium, humerus, and tarsometatarsus.
Provenience: Room 71, "two parrot skeletons were found in the southwest corner."
Elements: Major elements of one individual.
Remarks: From the collection of the American Museum of Natural History, New York. The fact that these macaw bones were associated with intrusive mammal bones, coupled with the fact that the character of the soil adhering to bones, identify the place where the skeletons were discovered as within a refuse deposit. Moreover, the presence of tracheal rings with the skeleton is evidence that this macaw was in the flesh when buried here. From archaeological field notes it might be possible to determine whether the macaw corpse was cast away and covered with trash or if the burial was a humanly made grave. The other parrot skeleton recorded with H/6452 could not be found.

H/6708 – Scarlet Macaw (*Ara macao*), Adolescent (1-3 years old). Species identification on premaxilla, basitemporal plate, basihyal-basibranchial, quadrate, manubrium, humerus, and tarsometatarsus.
Provenience: Room 78, in trash fill.
Elements: 153, representing the whole bird.
Remarks: Several bones were partly bleached with ashes, with powder still attached. Other remarks same as for H/6452.

H/6709 – Scarlet Macaw (*Ara macao*), Adolescent (1-3 years old). Species identification on premaxilla, basitemporal plate, basihyal-basibranchial, quadrate, manubrium, humerus, and tarsometatarsus.
Provenience: Room 78, in trash fill.
Elements: 252 whole or part elements representing the whole bird.
Remarks: The same as for H/6708.

USNM343578 – (No. 1 originally No. 285) – Scarlet Macaw (*Ara macao*), Adolescent (1-3 years). Species identification on premaxilla, basitemporal plate, and manubrium.
Provenience: (unknown).
Elements: premaxilla, cranium, seven vertebrae, right and left humeri, left femur, and sternum.
Pathology: Carinal margin of sternum broken over and healed.

USNM343576 – (No. 2 originally No. 286) – Scarlet Macaw (*Ara macao*), Newfledged (11-12 months). Species identification on premaxilla, basitemporal plate, manubrium, humerus, and tarsometatarsus.
Provenience: Room 249.
Elements: 37, representing whole bird.
Pathology: Right carpometacarpus, distal end of metacarpal III enlarged and spongy.

USNM343571 – (No. 3 originally No. 287) – Scarlet Macaw (*Ara macao*), Newfledged (11-12 months). Species identification on premaxilla, cranium, manubrium, humerus, and tarsometatarsus.
Provenience: (unknown).
Elements: 36, representing the whole bird.
Pathology: Ulnae roughened, with accretions on proximal head; accretions on metacarpal II of both carpometacarpi; carinal margin of sternum broken and healed.

USNM343572 – (No. 4 originally No. 288) – Scarlet Macaw (*Ara macao*), Newfledged (11-12 months). Species identification on premaxilla, basitemporal plate, manubrium, and humerus.
Provenience: (unknown).
Elements: 34, representing the whole bird.
Pathology: Carina of sternum damaged and healed; manubrium and sternal articulations of coracoids damaged and healed; overgrowth on postorbital processes of cranium.

USNM343573A – (No. 8 originally No. 474) – Scarlet Macaw (*Ara macao*), Adolescent (1-3 years). Species identification on basitemporal plate, manubrium, and humerus.
Provenience: Room 251.
Elements: 12, representing head, axial body and wings.

USNM343573B – (No. 8 originally No. 474) – Scarlet Macaw (*Ara macao*), Newfledged (11-12 months). Species identification on premaxilla.
Provenience: Room 251.
Elements: Major elements of one individual.
Pathology: All ulnae slightly roughened.

USNM343580 – (No. 13 originally No. 1200) – 3 Scarlet Macaw (*Ara macao*), all Newfledged (11-12 months).
Provenience: (unknown).
Elements: All major elements of one individual mixed with random bones of two others.
Pathology: All ulnae slightly roughened.

USNM343581 – (No. 14 originally No. 1201), Scarlet Macaw (*Ara macao*), Newfledged (11-12 months). Species identification on basitemporal plate, manubrium, humerus, and tarsometatarsus.
Provenience: (unknown).
Elements: 36, representing the whole bird.
Pathology: Roughened ulnae.

USNM343579 – (No. 16 originally No. 1259) – Scarlet Macaw (*Ara macao*), Newfledged (11-12 months). Species identification on premaxilla, cranium, humerus, and tarsometatarsus.
Provenience: (unknown).
Elements: 40, representing whole bird.
Pathology: Lump on left ulna, right radius bowed.
Remarks: Buried in the flesh (tracheal rings present).

USNM343574 – (No. 24 originally No. 503) – Scarlet Macaw (*Ara macao*), Newfledged (11-12 months). Species identification on premaxilla, cranium, humerus, and tarsometatarsus.
Provenience: (unknown).
Elements: 34, representing the whole bird.
Pathology: Ulnae slightly roughened.

USNM343575 – (No. 25 originally No. 508) – Scarlet Macaw (*Ara macao*), Newfledged (11-12 months). Species identification on premaxilla, humerus, and tarsometatarsus.
Provenience: (unknown).
Elements: 26, axial skeleton, right wing, and right leg.

USNM343583 – (No. 27 originally No. 668) – Scarlet Macaw (*Ara macao*), Newfledged (11-12 months). Species identification on premaxilla, cranium, and manubrium.
Provenience: Kiva J.
Elements: Premaxilla, cranium, mandible, and sternum.

Comments. At the time Pepper conducted his investigation, there was no promise that some day wood used in building a pueblo would provide dates in our calendar and thus make possible time correlations in interpreting prehistoric events. By the time Judd conducted his investigations, promise had almost become certainty through the efforts of Douglass (1929), who demonstrated that wood specimens could be relied upon to provide correct calendar dates for the growth of trees.

In the case of Pueblo Bonito, only one of the 98 dated wood specimens came from a room in which macaw remains were found. Specimen JPB-37, from Pilaster 5, Kiva J, yielded a date of 925-1080vv (Bannister 1965: 180). The latest tree-ring date from the Pueblo is 1126 from Specimen JPB-17 (Bannister 1965: 183). Macaws then must have been in the pueblo about 1100 or shortly before.

Pueblo del Arroyo

Situated on the north bank of the Chaco River, Pueblo del Arroyo is less than a quarter mile from Pueblo Bonito. Pueblo del Arroyo may have stood four stories high and may have housed as many as 800 people (Hewett 1936). Beginning in 1923 Judd conducted excavations here for the National Geographic Society. Of 45 dated tree-ring specimens, the latest date is 1117 ± (JPB-148 collected by Judd; Bannister 1965: 188).

Bones Recovered by Judd

Judd (1959: 127) found three "articulated" macaw skeletons on a shallow accumulation of sand in Room 63, and an incomplete skeleton in Room 44. Two of the skeletons from Room 63 were available for restudy.

USNM344360 – (No. 400 originally No. 548) – Scarlet Macaw (*Ara macao*), Newfledged (11-12 months). Species identification on premaxilla, basitemporal plate, basihyal-basibranchial, manubrium, humerus, and tarsometatarsus.
Provenience: Room 63.
Elements: 51, representing entire bird.

USNM344359 – (No. 402 originally No. 549) – Scarlet Macaw (*Ara macao*), Newfledged (11-12 months). Species identification on premaxilla, basitemporal plate, quadrate, basihyal-basibranchial, humerus, and tarsometatarsus.
Provenience: Room 63.
Elements: 89, representing entire bird.
Pathology: Ulnae slightly roughened; round hole in distal end of left ulna – appears to be result of abcess, healing.
Remarks: Buried in the flesh (tracheal rings present).

Bones Recovered by Vivian and Rixey

In 1949 Vivian and Rixey unearthed an "articulated" macaw in Room 50. The presence of tracheal rings is evidence that the macaw was buried in the flesh.

C5 – Scarlet Macaw (*Ara macao*), Newfledged (11-12 months old).
Provenience: Room 50.
Elements: 54, representing head, neck and tail; tongue and trachea; axial body, wings, and legs.
Remarks: Identified on the presence of six diagnostic traits. Buried in the flesh (tracheal rings present).

Kin Kletso

On the north side of the Chaco Wash, about one-half mile below Pueblo Bonito is Kin Kletso, another multistoried pueblo, with building dates ranging from 1059 to 1124. Two dated pieces of charcoal, CKK-27-1 and CKK-26-1, provided outside dates of 1171 and 1178; these were from a firepit and an ash layer in Room 24 (Bannister 1965: 171). One macaw specimen (Provenience 6/358) was recovered in 1951-53 by Vivian (Vivian and Mathews 1965).

C163 – Scarlet Macaw (*Ara macao*), Adolescent (1-3 years). Identified on two diagnostic characters.
Remarks: These two bones, a skull and associated premaxillary may have been the remains of a stuffed macaw.

Puerco River (West)

The Puerco River rising near the heart of the Chaco Canyon Area flows southwest to Holbrook where it joins the Little Colorado River on its way northwest to the Colorado River at the Grand Canyon. Although seemingly offering no great physical obstruction, in prehistoric time the arc formed by these two streams appears as a cultural or psychological barrier between the Anasazi toward the north and the Mogollon and Sinagua toward the south and west. These cultural areas are easily distinguished because the Anasazi employed the reducing atmosphere as a ceramic technique whereas peoples south of the arc fired their pottery in the oxidizing atmosphere which resulted in differently colored vessels (Hargrave 1932: 7; Colton and Hargrave 1937: 8). This situation persisted until about 1200 when the river "barrier" became broken along the lower Puerco and the Little Colorado River; at least Mogollon-like culture traits then moved north in mass over a deep broad front through abandoned Anasazi territory from Houck to Winslow, thence to the north escarpment of the Little Colorado Valley. Soon after 1300, as determined principally from ceramic techniques and pottery types (Colton and Hargrave 1937), this Mogollon-derived culture reached as far as the Jeddito Valley and the Hopi Mesas (Hargrave 1931a, 1931b, 1932, 1937).

Houck K, NA8440

A northward movement of southern influences at any date may have had a bearing on the macaw problem, and may have done so in the case of Houck K, near the junction of Black Creek with the Puerco River, where a macaw bone was found.

This site was excavated by the Museum of Northern Arizona in 1962 under the direction of Alan P. Olson. He reports that it is "a Chacoan site which, at least in architecture, is intrusive into the Houck Area." Kidder (1924: 56, 57) reports Chaco sites along Black Creek, about 16 miles south of Fort Defiance. Black Creek joins the Puerco at Houck. No tree-ring dates were recovered from Room 1, where the macaw bone was found in floor fill. However, in the assemblage of pottery types recovered were sherds of one painted pottery type of the Chaco Complex (Anasazi), that is, Gallup Black-on-white which predates 1100.

Six non-Chacoan pottery types recovered from Room 1, floor fill, were identified by Olson as:

Puerco Black-on-red	Post	950 - 1150
St. Johns Polychrome		1100 - 1200
Walnut Black-on-white		1100 - 1250
Pinto Polychrome	About(?)	1150 - 1250
Houck Polychrome	About	1200 - 1250
Queriño Polychrome	About	1250 - 1350

Of these six pottery types, three (St. Johns Polychrome, Walnut Black-on-white, and Pinto Polychrome) were traded from other areas, and three (Puerco Black-on-red, Houck Polychrome, and Queriño Polychrome) were of the Houck Series (Colton and Hargrave 1937: 118) and are indigenous to the area.

Because of the occurrence of pottery types of two cultural complexes associated in trash fill, the problem of determining to which end of the time-sequence this macaw bone belongs has not been solved by Olson, but he suggests that the macaw most likely can be placed in the mid-1200s.

Bones Recovered

A0804 — Scarlet Macaw (*Ara macao*), Newfledged (11-12 months old) or older.
Elements: One (the sternum) representing the axial body.
Remarks: Identified by the diagnostic features of the manubrial spine (dorsal view). With well-defined muscle attachments, this bone may be too old for Newfledged.

Kayenta Area

The Tsegi Canyons of the Kayenta Area include part of the southern drainage of the San Juan River system (Hargrave 1934). These canyons from the headwaters of Laguna Creek, which is the west fork of Chinle Creek that flows to the San Juan River, whose system embraces most of the territory occupied by the Anasazi Culture south of the Colorado River (Hargrave 1935: 22). Within the system of these drainages were concentrations of pueblos manifesting cultural kinships which tied the Kayenta Area to other areas of the same cultural tradition, such as Mesa Verde and Chaco Canyon.

There are small and large cliff pueblos in the Tsegi Canyons, some in pine and fir-studded settings high in faces of canyon walls, often with deciduous-bordered streambeds below. This picturesque area (in the 1960s) is still a foot-and-horseback country of many miles of rough canyons and mesas. The very roughness of the country and steepness of the cliff walls have discouraged investigations, which, nevertheless, were begun in 1907 and since then have been made at intervals into the early 1930s (Hargrave 1935: 11-17; Beals, Brainerd, and Smith 1945). To my knowledge the only newly discovered cave site is Woodchuck Cave, a Basketmaker cave at the head of Water Lily Canyon, excavated in 1934 (Hargrave 1935; Lockett and Hargrave 1953).

Temporal correlations from pottery types recovered by the Rainbow Bridge–Monument Valley Expeditions and recent tree-ring studies indicate that the plateau sites were abandoned a little before or about 1250 (Dean 1964: 6). Near this time an increase in building took place in the Tsegi Canyons, as shown at several cliff pueblos where early available tree-ring dates range from 1057 at "Ladderhouse" (NA2543; in quotes to distinguish it from similarly named sites) to 1124 at Lenaki (NA2630; Smiley 1951: 15) and to 1250 at Long House (Dean 1964: 6). It is reasonable to assume, I think, that while small family-size dwellings were scattered in open country outside of canyon caves, that is, pre-1125, these same cliff caves were occupied by similar "family" groups of the same cultural level as those at "Ladderhouse" and Lenaki. Two or three small house units could easily be obliterated or submerged by a sudden addition of more buildings by newly arrived small-unit dwellers, the original two or three units serving as a nucleus for a full-fledged pueblo.

It is not clear what took place in the canyons during the next 150 years, but there were large pueblos outside the canyons at 1125. Then, rather suddenly there was a spurt of active building in some caves in the early 1270s (Smiley 1951). This is the way Dean (1964: 23) summarizes this situation at Kiet Siel at this time: "The major conclusion to be drawn. . .is that Kiet Siel at this time was largely built in one burst of building between 1272 and 1275."

From Smiley's dates the Tsegi Canyons must have been abandoned at about the same time, soon after the 1270 building spurt, since the latest date is 1289 (repair) from Twin Caves (NA2536). According to Dean (1964: 2, 12, 32), the latest date from Kiet Siel is 1286, after which date "there was a gradual decline in population as people died off or as families left villages, until, perhaps around 1300 or a little later, the remaining population moved away abandoning Kiet Siel forever."

Kiet Siel Pueblo

Kiet Siel Pueblo, in Navajo National Monument, is the largest cliff pueblo in Arizona. Kiet Siel was first excavated by Richard Wetherill in 1897 for the Hyde Brothers, and in 1934 the site was partially excavated and stabilized with funds provided by the Civil Works Administration (Wetherill 1934; Hargrave 1935). It was during this work that a macaw burial was found. Although few archaeological bones were saved in those days, the skeleton obviously being an exceptional item was carefully preserved as an artifact (695/NA2519.276). Later a single macaw bone was found.

Macaw burials and isolated skulls and bills are readily identified by excavators, but random bones seldom are identified as macaw. Thus in masses of miscellaneous macaw bones from refuse accumulations, many scattered bones or fragments do not display any recognizable species character, although generic characters are recognizable; these bones are identified and recorded as Macaw (*Ara* sp.).

Bones Recovered

A052 – Macaw (*Ara* sp.), Newfledged (11-12 months old).
Provenience: (unknown).
Elements: Right tibiotarsus.
Remarks: No species characters are preserved on this specimen, but in size and proportion it falls within the range of all Scarlet Macaws discussed herein.

A0437 – Macaw (*Ara* sp.), Newfledged (11-12 months old).
Provenience: Two feet deep in the trash of Room 49.
Elements: 20, representing the head, axial body, wings, and legs.
Pathology: A pathological ulna is slightly roughened (Fig. 17b), and a left suborbital arch had fractured obliquely, broken ends healing without reuniting.
Remarks: This macaw had a piece of skin with gray down still attached to the bend of the wing, indicating that it had probably been denuded of colored feathers before burial.

Comments. The macaw skeleton numbered A0437 is exceptional in several ways that can best be shown in a tabulation of species characters as compared to *Ara militaris* and *Ara macao*. This

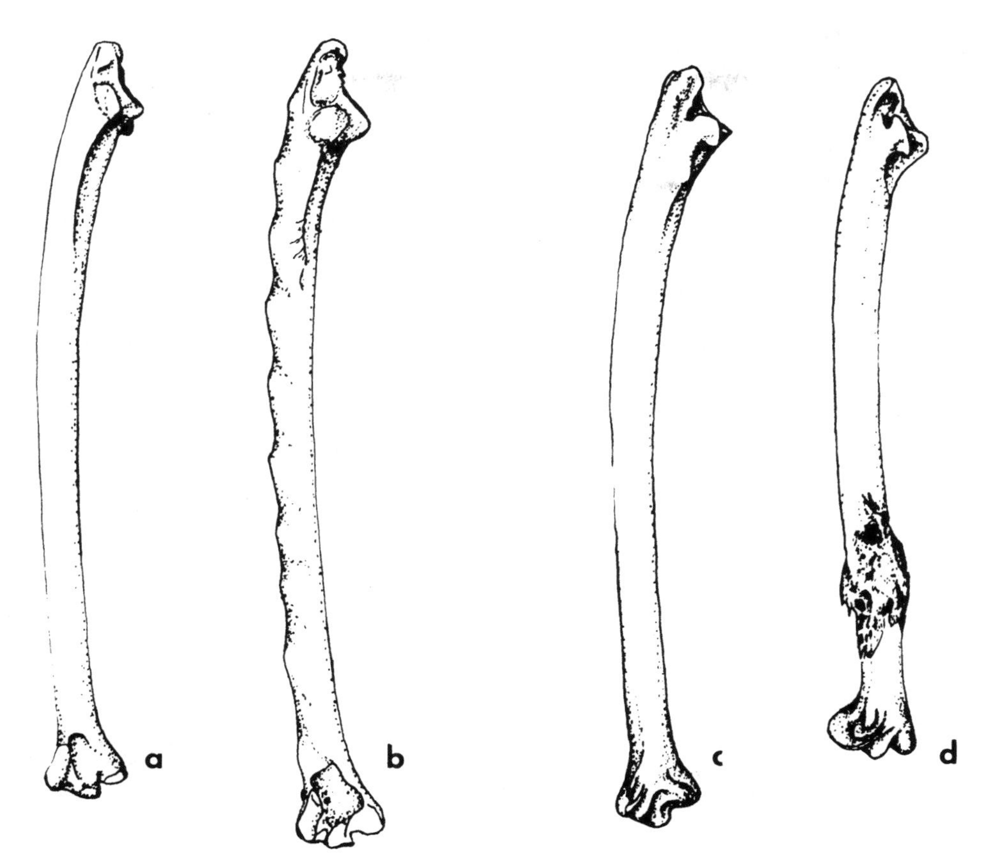

Fig. 17. Normal and pathological bone specimens. *a, Ara macao* (H2099), normal left ulna; *b, Ara macao* (A0537), "roughened" left ulna; *c, Ara macao* (H2159), normal right ulna; *d, Ara militaris* (H2440), fractured right ulna healed with malunion including shortening and rotation of shaft. (Scale: X 1)

tabulation is based on 10 characters that are definitive as to species, 7 of which are manifested on skeletal elements of *Ara militaris* and *Ara macao*. These characters are compared to homologous characters of No. A0437, with the result that No. A0437 has (1) one character in common with *Ara militaris*, (2) three characters in common with *Ara macao*, and (3) six characters unlike either *Ara militaris* or *Ara macao*.

In addition to the 10 definitive characters used in this comparison, by inspection alone from its relative size and differently proportioned bones, A0437 is a small, stocky, big-headed, and long-billed macaw, and is thus unlike any specimen I have seen of either *Ara militaris* or *Ara macao*. Several possibilities for a logical solution for this problem exist.

From the diagnostic skeletal character differences between *Ara militaris* and *Ara macao*, as enumerated previously and listed in Table 3, the Macaw skeleton A0437 may be the remains of some other described and named Macaw not included in this study.

It is even conceivable that A0437 may be the remains of an unknown Macaw, in which case it would be a new species that appropriately should be named *Ara wetmorei* in recognition of Alexander Wetmore's contribution in identifying numerous archaeological bird bones from southwestern ruins.

The macaw skeleton A0437 was found buried two feet deep in trash north of Room 49. As this burial was between the retaining wall of Court 63 (tree-ring date 1274; Dean 1964: 13) and the wall protecting Room 74 (tree-ring date AD 1275; Dean 1964: 15), it is probable that this trash fill was deposited shortly after 1274-75. Thus, as deduced from Dean, the macaw most probably was buried here between 1275 and 1286.

SINAGUA CULTURE AREA

On the north, the territory of the Sinagua abuts the territory of the Anasazi along the Little Colorado River to Winslow, where it roughly borders Mogollon territory on the east; on the west, Sinagua territory extends to the west side of the Verde Valley, thence south along the eastern border of Hohokam territory (Fig. 15).

Ceramic cultural indicators for the area are (1) applied decoration–rarely painted (white), (2) surface color–normally shades of brown, (3) surface finish of utility vessels–noncorrugated, (4) firing atmosphere–oxidizing, and (5) method of thinning vessel walls–paddle and anvil.

San Francisco Mountain Area, Arizona

The archaeology of the San Francisco Mountain area has been studied in detail by the Museum of Northern Arizona. The outstanding result of several years of investigations here has been the recognition of the Sinagua Culture (Colton 1946).

The entire collection of macaw bones from the San Francisco Mountains has recently been restudied as part of this project, and some different assessments and corrections have been made and recorded to conform with new data.

Wupatki, NA405

Wupatki is one of the largest pueblo ruins in the San Francisco Mountain area and is located on the northern perimeter of the Sinagua cultural area, only 7 miles southwest of the Little Colorado River. It is the major ruin in Wupatki National Monument.

The first published recovery of macaw remains from prehistoric Indian sites in Arizona appears to be three skeletons dug up at Wupatki Pueblo by members of the Museum of Northern Arizona Archaeological Expedition 1933, under my supervision (Hargrave 1933: 26). These carcasses were wrapped in rush matting for burial, as shown by imprints in the soil, which denotes human care in the burial of the dead.

Wupatki Pueblo is constructed of masonry. There are about 70 ground-floor rooms, and in some sections two additional upper stories. In all, 37 rooms have been excavated by the Museum of Northern Arizona and the National Park Service. Most remains of macaws came from these excavations, although a few specimens were later recovered as salvage by stabilization crews of the National Park Service.

Although the pueblo was built in the open, thus exposed to the elements, the hot sun and dry air of the Little Colorado Valley preserved numerous construction timbers from which 49 reliable tree-ring dates have been recorded.

These dates range from 1073 to 1205, but, from other studies, Colton (1946: 58-63) thinks that Wupatki as a pueblo dates from 1120 or 1130 to between 1200 and 1225.

Bones Recovered

A0.50 – Macaw (*Ara* sp.). No age or species characters present.
Provenience: Trash fill, 1933.
Elements: Distal head, left ulna.
Pathology: Shaft roughened.

A0.79 – Macaw (*Ara* sp.). No age or species characters present.
Provenience: Trash heap.
Elements: Distal half, left tarsometatarsus.

A0.87 – Macaw (*Ara* sp.). No age or species characters present.
Provenience: 54.37.
Elements: Distal 2/3, left ulna.

A0.160 – Macaw (*Ara* sp.). No age or species characters present.
Provenience: Room 41.
Elements: Half of radius; 1 rib.

A0.183 – Macaw (*Ara* sp.). No age or species characters present.
Provenience: Not given.
Elements: Part of the skull.

A0.285 – Macaw (*Ara* sp.). No age or species characters present.
Provenience: Not given.
Elements: Left palatine.

A0.322 – Macaw (*Ara* sp.). No age or species characters present.
Provenience: Miscellaneous trash.
Elements: Left ulna, proximal end missing.
Pathology: Shaft roughened.

A0.323 – Macaw (*Ara* sp.), Newfledged (11-12 months old, or older).
Provenience: Room 43, trash fill.
Elements: Right carpometacarpus.

A0.324 – Macaw (*Ara* sp.). No age or species characters present.
Provenience: 54.37.
Elements: Pelvis.

A0.529 – Macaw (*Ara* sp.). No age or species characters present.
Provenience: Room 48.
Elements: Right coracoid (chipped).

A0.531 – Macaw (*Ara* sp.), Newfledged (11-12 months old).
Provenience: M. 177.
Elements: Mandible, part furcula, distal end of right tibiotarsus.

A0.541 – Macaw (*Ara* sp.), Newfledged (11-12 months old). No species characters.
Provenience: B.8.1
Elements: 35, representing head, neck trachea, axial body, right wing, and legs.
Pathology: Right femur, proximal head broken and healed with accretions.
Remarks: Burial (trachea present).

A0.555 – Macaw (*Ara* sp.). Newfledged (11-12 months old). No species characters.
Provenience: General screenings.
Elements: Distal end, right tibiotarsus.

A0.556 – Macaw (*Ara* sp.). No age or species characters present.
Provenience: General screenings.
Elements: Right ulna.

A0.752 – Macaw (*Ara* sp.). No age or species characters.
Provenience: Room 62.
Elements: One rib.

A0.753 – Macaw (*Ara* sp.). No age or species characters.
Provenience: Screening May 1959
Elements: Left coracoid (broken).

A0.755 – Macaw (*Ara* sp.). No age or species characters.
Provenience: Room 41.
Elements: Part palatine.

A0.757 – Macaw (*Ara* sp.), Newfledged (11-12 months old). No species characters.
Provenience: Screening January 1940.
Elements: 6, representing the neck and tail, left wing, and right leg and foot.
Pathology: Shaft of ulna roughened.

A0.766 – Macaw (*Ara* sp.). No age or species characters.
Provenience: Screening May 1939.
Elements: Part pelvis and one caudal vertebra.

A0.80 – Scarlet Macaw (*Ara macao*). No age characters.
Provenience: Screening trash heap.
Elements: Right humerus (proximal 2/3) with diagnostic species characters.

A0.81 – Scarlet Macaw (*Ara macao*). No age characters.
Provenience: Trash heap.
Elements: Right humerus, proximal 2/3, with diagnostic species characters.

A0.325 – Scarlet Macaw (*Ara macao*). No age characters.
Provenience: Miscellaneous trash.
Elements: Right humerus, with diagnostic species characters.

A0.528 – Scarlet Macaw (*Ara macao*). No age character. Species identification based on head characters and the humerus.
Provenience: Room 73 (has been dug twice).
Elements: Skull, mandible, premaxilla, right coracoid furcula, right humerus, and left femur.
Pathology: Accretions on head of humerus.
Remarks: A0.528 includes old catalogue numbers A & B and W99.

A0.533 – Scarlet Macaw (*Ara macao*). No age character.
Provenience: Room 41.
Elements: Left humerus and left ulna.
Pathology: Lump on humerus as from healed injury.

A0.534 – Scarlet Macaw (*Ara macao*), Immature (4-11 months). Identification made on three diagnostic characters, basitemporal plate, quadrate, and premaxilla.
Provenience: Room 67.
Elements: 9, representing head and axial body (sternum and pelvis).

A0.535 – Scarlet Macaw (*Ara macao*), Immature (4-11 months). Species identification based on quadrate, humerus, and tarsometatarsus; age determined on character of the tibiotarsus.
Provenience: Room 48.
Elements: 25, representing major segments.
Remarks: Probably a burial.

A0.536 – Scarlet Macaw (*Ara macao*), Newfledged (11-12 months old). Species identification on character of humerus.
Provenience: Not given.
Elements: 16, representing the head, neck trachea, axial body and wings.
Remarks: Tracheal rings and bones from most segments determine this was a burial made in the flesh.

A0.537 – Scarlet Macaw (*Ara Macao*), Newfledged (11-12 months old). Species identified on three diagnostic characters.
Provenience: Room 63.
Elements: 39, representing head, neck and tail, tongue and trachea, axial body, wings, and legs.
Pathology: Ulnae roughened, one vertebra fused to pelvis.
Remarks: Burial in the flesh (tracheal ring present).

A0.538 – Scarlet Macaw (*Ara macao*). Newfledged (11-12 months old). Species identified on character of the premaxilla, basihyal-basibranchial.
Provenience: Room 43.

Elements: 46, representing head, neck, tongue, trachea, axial body, wings, and legs.
Pathology: Ulna roughened, lump on shaft of right femur as from healed injury.
Remarks: Burial (trachea present).

A0.539 – Scarlet Macaw (*Ara macao*), Newfledged (11-12 months old). Species identified on premaxilla, basitemporal plate, quadrate, basihyal-basibranchial, and humerus.
Provenience: Not given.
Elements: 40, head, neck, tail, tongue and trachea, axial body, wings, and legs.
Pathology: Ulnae roughened, raised areas on shaft of right humerus, and accretions on distal head of right femur.
Remarks: Buried in the flesh (trachea present).

A0.540 – Scarlet Macaw (*Ara macao*), Newfledged (11-12 months old). Species identified on seven diagnostic characters.
Provenience: B.10.1.
Elements: 178, representing body.
Pathology: Roughened ulnae; left humerus, proximal head, palmar surface, crushed and healed.
Remarks: Buried in the flesh (tracheal rings present).
Note: A0.540 includes bones numbered A0.521 and A0.450.

A0.542 – Scarlet Macaw (*Ara macao*), Immature (4-11 months). Species identification made on premaxilla, quadrate, basihyal-basibranchial and manubrium.
Provenience: B.11.1.
Elements: 135, nearly complete body.
Pathology: Tip of right ischium broken and healed at an angle.
Remarks: Buried in the flesh (tracheal rings present).

A0.544 – Scarlet Macaw (*Ara macao*), Immature (4-11 months old). Species identification made on character of basitemporal plate.
Provenience: Room 35C.
Elements: One, cranium.

A0.547 – Scarlet Macaw (*Ara macao*). Newfledged (11-12 months old). Species identification made on character of basitemporal plate.
Provenience: Not given.
Elements: 2.

A0.548 – Scarlet Macaw (*Ara macao*), Breeding (4-? years). Species identification made on character of basitemporal plate; age on fenestrae.
Provenience: Room 35 A-17.
Elements: One, cranium.

A0.549 – Scarlet Macaw (*Ara macao*), Newfledged (11-12 months). Identified to species on basitemporal plate.
Provenience: Not given.
Elements: Skull only.

A0.764 – Scarlet Macaw (*Ara macao*). Species identification made on basihyal-basibranchial and tarsometatarsus; age unknown.
Provenience: Screening May 1939.
Elements: Representing the neck, tongue, axial body, left wing, and right leg.
Note: Included former numbers A0. 763, 764, 765, 768, 769, 770, 771, 772, 774, 775, 776, 778, 779 and 780.

A0.773 – Scarlet Macaw (*Ara macao*). Species identification made on tarsometatarsus; age unknown.
Provenience: Screening May 1939.
Elements: 2, right tarsometatarsus and right ungual.
Note: Formerly included in A0.777.

W50 – Scarlet Macaw (*Ara macao*), Newfledged (11-12 months old). Species identified on five diagnostic characters.
Provenience: Room 80-81, fill.
Elements: 65, head, neck and tail, trachea, axial body, left wing and legs.
Pathology: Left ulna roughened, lump on shaft of left humerus as from healed injury.
Remarks: Buried in the flesh (tracheal rings present).

W84 – Scarlet Macaw (*Ara macao*), Newfledged (11-12 months). Species identified on premaxilla and basitemporal plate.
Provenience: Room 80-81, fill.
Elements: Premaxillary, skull, and mandible.

W105 – Scarlet Macaw (*Ara macao*), Newfledged (11-12 months old). Species identified on premaxilla, basitemporal plate, quadrate, manubrium, humerus, and tarsometatarsus.
Provenience: Room 81, S. Side.
Elements: 32, representing the head, neck, axial body, wings, legs, and tail.

Comments. A total of 685 macaw bones has been recorded from Wupatki Pueblo, representing an estimated number of 41 individuals, 22 of which are Scarlet Macaws (*Ara macao*). The remaining 19 individuals cannot be identified to species because they are lacking in diagnostic species characters.

Nalakihu, NA358

Nalakihu is a ground-floor masonry pueblo of ten rooms near the north base of a large, conspicuous pueblo (NA355) called the Citadel, in Wupatki National Monument. The Citadel and Nalakihu are so close to each other that they might well be one site, as indicated also by closeness in tree-ring dates. On the basis of tree-rings and ceramic types, Colton (1946: 52, 53) has reviewed the ages of the Citadel and Nalakihu. From these data I agree that the occupation range of the Citadel should fall between 1125 and 1200, but in my opinion the occupation range of Nalakihu is between 1130 (Colton's ceramic date) and about 1200 (McGregor's four bark dates of 1185). The abandonment dates of Wupatki, the Citadel, and Nalakihu then probably occurred at about the same time. Some pottery specimens could be late intrusives at the sites.

Bones Recovered

A0.326 a,b – Macaw (*Ara* sp.), Adolescent or older (1-3 years). Age stage assigned from characters of the supratendinal bridge of the distal end of the carpometacarpus.
Provenience: Pit 8 (King 1949: 57).
Elements: A mated pair—right and left carpometacarpus.
Remarks: In the early 1930s I mistakenly identified it as *Ara militaris* (King 1949: 141). I now know that the development of the supratendinal bridge is an age and not a species character.

Comments. A not uncommon practice in the identification of archaeological bones has been the use of "identified bones" recovered and studied from various sites as comparative material in identifying other unknown bones. This is a bad practice, as illustrated by my study of these two carpometacarpi (A0.326 a, b). In the 1930s I did not have modern skeletons of the Military Macaw (*Ara militaris*) or the Scarlet Macaw (*Ara macao*) for comparison, but I did have archaeological bones which had been identified as the Scarlet Macaw. When I found I had macaw bones unlike the bones of the archaeological Scarlet Macaws on hand, and believing only one other species of macaw, the Military Macaw, could be involved in my problem, I unhesitatingly identified and named the pair of carpometacarpi as the Military Macaw only because they were not like the carpometacarpi with which I compared them. I know I was wrong and admit that I cannot distinguish the carpometacarpus of the Military Macaw from the carpometacarpus of the Scarlet Macaw. I therefore retract my identification of A0.326 a, b as the Military Macaw (*Ara militaris*) and identify it simply as Macaw (*Ara* sp.).

The carpometacarpi were found in trash fill of an abandoned oven. Trash consisted of quantities of bones and sherds and a bowl of Sunset Red (King 1949: 56, 57). Sunset Red is a pottery type with a time range from about 1050 to about 1200 (Colton and Hargrave 1937: 163).

Growth Rate of Carpometacarpus. The growth rate of the carpometacarpus was not included in the presentation of the growth rate of bones of the genus *Ara* because of insufficient comparative material of modern macaws of known age at death. Because the carpometacarpi numbered A0.326 a, b were misidentified, and because, unfortunately, the species identification was published, the available data are now presented here and illustrated in Figure 18.

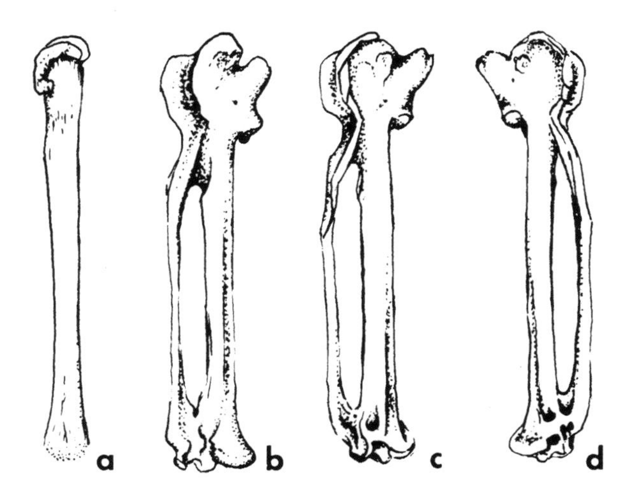

Fig. 18. Formation of supratendinal bridges of the carpometacarpus. *a, Ara chloroptera* (H2449) 8 weeks, distal end not yet calcified, metacarpal III not yet joined to metacarpal II; *b, Ara macao* (D2211) Immature (4-11 months), both supratendinal bridges forming but not yet complete; *c, Ara militaris* (H2440) Newfledged (11-12 months), carpometacarpus, internal view, showing one supratendinal bridge complete; *d,* carpometacarpus, internal view, from previous individual (H2440), showing both supratendinal bridges completed. (Scale: X 1)

This seeming discrepancy between the ossification of corresponding elements within one individual, or between several individuals of a species, is not unusual among large long-lived birds such as macaws.

Ridge Ruin, NA1785

Ridge Ruin, about 20 miles northeast of Flagstaff (and in Township 21 North, Range 10 East, Section 18), was excavated in 1938 by John McGregor (1941) of the Museum of Northern Arizona. While excavating Room 13, McGregor (1943: 270-98) found a human burial with a fantastic assortment of funeral offerings buried beneath the floor. A macaw burial was found above the floor in the northwest corner.

Bones Recovered

A0.467 – Scarlet Macaw (*Ara macao*), Newfledged (11-12 months old). Identified to species on characters of the premaxillary and tarsometatarsus.
Provenience: Room 13.
Elements: 41, head, neck, trachea, axial body, wings and legs.
Pathology: Greatly roughened ulnae.
Remarks: Buried in the flesh (trachea present). Originally identified by Howard (McGregor 1941: 6, 258); recently verified by Hargrave.

Comments. Although this macaw (A0.467) was buried in Room 13, there seems to be no connection to a human burial beneath the floor of the room, except that the human was buried before the macaw was buried. This is evidenced by field notes on the box in which the macaw bones originally were packed. These notes read "buried on [not in] cinders taken from burial pit and on east side of room." The grave for the human corpse was dug through the room floor into a volcanic cinder subsoil. The problem then is to determine an approximate date when each of these two bodies was buried.

Since the human burial was the older, an analysis was made of pottery vessels from the grave, of which McGregor recovered 25 (McGregor 1943: 270). Six pottery types represented–Citadel Polychrome, Tusazan Polychrome, Turkey Hill Red, Sunset Red, Flagstaff Black-on-white, and Walnut Black-on-white – gave ceramic dates (based on Colton and Hargrave 1937) from which I have determined that the human was buried about 1150. No evidence, was found as to just when the macaw was buried; but after Room 13 was converted to a burial chamber, it was further used as a trash pit. When McGregor began excavating Room 13 it had been completely filled with refuse from Ridge Ruin Pueblo above it; the pottery types–Citadel Polychrome, Flagstaff Black-on-white, and Walnut Black-on-white–were within the time given by tree-ring dates ranging from 1085 to 1173.

Two random macaw bones were recovered from the trash fill of Room 13 above the macaw burial A0.467.

A0.494 – Macaw (*Ara* sp.), Immature (4-11 months old). No species character present.
Provenience: Room 13, trash fill.
Elements: Right femur (proximal 4/5th); head chipped.
Remarks: This short and stocky bone was identified as "Immature" because of faintness of the intermuscular line and the porosity (texture) of the proximal end. It cannot be older.

A0.497 – Scarlet Macaw (*Ara macao*), Newfledged or older. Identified to species on character of premaxilla.
Provenience: Room 13, trash fill.
Elements: Premaxilla (part of). Identified by Howard; verified by Hargrave.

Comments. Since the two random bones were found well above the floor of Room 13 in trash from the pueblo, these bones, as trash, date later than 1150 but probably earlier than 1200, a date established as a local cultural division (Colton 1946: 17).

A0.468 – Scarlet Macaw (*Ara macao*), Immature (4-11 months). Identified to species on character of proximal foramen of tarsometatarsus.
Provenience: Trash mound test trench.
Elements: Right tarsometatarsus. *Age Characters* (*1*) proximal articular surface, porous; (*2*) proximal outer foramen not yet fully constricted; (*3*) bridge across opening in intercondylar area not yet formed (Proximal view).
Remarks: The most disappointing of all the trash mounds sectioned (McGregor 1941: 25). Ceramic date about the same as A0.497.

Winona Village, NA3644

Winona Village is about 15 miles east of Flagstaff, in Section 14, Township 21 North, Range 9 East. It is one of the group of pithouses excavated by McGregor (1941: 116) in 1939.

Bones Recovered

A0.426 – Macaw (*Ara* sp.). Age unknown. Not identifiable to species.
Provenience: From fill on upper floor?
Elements: One left femur of normal size. Chipped in several places.

Comments. Information as to the exact provenience of this bone (A0.426) seems to have been misplaced; but since the record definitely indicates a structure with a trash fill, there seems to be little doubt that the bone came from the trash accumulation on top of the latest of two floors. Although there are no tree-ring dates from the structure, Colton (1946: 231) has listed the pottery types represented—Winona Brown, Winona Smudged, Winona Corrugated, and Turkey Hill Red. From an analysis of the time range of four indigenous companion types, as given by Colton and Hargrave (1937: 52, 55, 162, 165) a close ceramic date of 1150 for the occurrence of this macaw bone is given.

Pollock Site, NA4317

The Pollock Site, about 40 miles east of Flagstaff, at the mouth of Kinnikinnick Canyon, on Anderson Mesa, was excavated by McGregor of the University of Illinois Archaeological Field School in 1953 and 1955 (Bannister, Hannah, and Robinson 1966: 22).

Bones Recovered

A0.710 – Scarlet Macaw (*Ara macao*), Newfledged (11-12 months old or older); identified to species on character of humerus.
Provenience: From Room 5.
Elements: 10, skull broken, sternum fragment, left scapula, right and left carpometacarpus, and left scapholunar.
Pathology: distal head of left coracoid fused with scapholunar.

Comments. This macaw (A0.710) came from Room 5. The recovery of related bones from a single individual must indicate a "burial" in the sense that a grave was prepared or the carcass was disposed of in room fill. We did not know which. No analysis of associated potsherds is available for temporal association. Although there are 23 tree-ring dates from the pueblo, none relate to Room 5, from within which the macaw bones came, so we cannot even determine a pre-disposal date.

There is an outside range of tree-ring dates from 1150 (F-6054B) to 1286 (F-6018), but the only two tree-ring dates considered fairly representative of cutting were 1284 (F-6024) and 1286 (F-6018; Bannister, Hannah, and Robinson 1966: 22-3). Therefore, the macaw bones were probably deposited after this building period from 1284 to 1286.

Verde River Valley

The western north-south border of the Sinagua area is indicated by the western tributary heads of the Verde River, the principal tributaries of which rise easterly under the Mogollon Rim. The Verde River flows south into the Salt River, which in turn drains the southeastern Sinagua area. Within the drainage of the Verde River are numerous abandoned Indian dwellings, three of which have been recorded as sites of macaw bones.

Montezuma Castle

Montezuma Castle is a spectacular pueblo built into the cliff face on the north side of Beaver Creek, a northerly draining tributary of the Verde River in Arizona. Other structures once were built in the same cliff face. These ruins and other archaeological features are within the boundary of Montezuma National Monument.

About 100 yards southwest of Montezuma Castle are the remains of a fallen cliff dwelling, Castle A (NA1278A), which at one time was twice the size of Montezuma Castle (Jackson and Van Valkenburgh 1954: 4). Several macaw bones were found in trash fill of this ruin.

Bones Recovered

J514a – Scarlet Macaw (*Ara macao*), Newfledged (11-12 months old). Identified to species on character of humerus.
Provenience: Lot of bones J514, Castle A, 1934 CWA excavation.
Elements: 5 from one individual—right and left humerus, left ulna, right carpometacarpus, and right tibiotarsus.
Pathology: Injury near distal head of carpometacarpus.

Comments. A miscellaneous lot of macaw bones, such as this one, might indicate bones from a disturbed macaw burial.

The ceramic dates for the earliest sequence at Montezuma Castle run from about 1125 through 1200, and may even extend to a later date (Jackson and Van Valkenburgh 1954: 4, 6, Table 3).

Jackson Homestead

About one mile south of Montezuma Castle, on the old Jackson Homestead are numerous prehistoric remains. Earl Jackson has informed me that in the mid-1920s he and his mother salvaged a macaw burial from beneath a narrow limestone ledge near where pothunters had removed several human baby burials. The macaw bones were wrapped in reed matting. With the macaw bones was a small crudely formed pottery bowl (J136). Presumably this vessel was a burial offering, as were other vessels found with the human burials, but I have never before seen a macaw burial with funeral offerings. I identify the vessel as Verde Brown, an indigenous type of long temporal range.

Bones Recovered

J121 – Scarlet Macaw (*Ara macao*), Adolescent (1-3 years old). Identified to species on skull characters.
Provenience: MCNM Accession 21. Burial under limestone ledge is located on map sketch in file at Montezuma National Monument where the bones are deposited.
Elements: Skull and mandible.

Comments. An important consideration in regard to this macaw burial is the few bones represented. Jackson's recollection is hazy about this, and he is not certain whether other bones were present. However, he was impressed by the absence of the premaxillary (upper part of the bill), only the mandible (the lower part of the beak) and the skull being present. The firm condition of these salvaged bones, the fact that no other large odd-shaped bones were saved, and the presence of perishable material in the grave, are strong evidence that the bones recorded here constituted the buried "skeleton." The fact that this buried macaw head was found in a child's "graveyard" suggests that it may have been a prized but broken toy.

Tuzigoot Pueblo

Tuzigoot Pueblo is situated on a ridge about 120 feet above the Verde River near Clarkdale, Arizona. It is in Tuzigoot National Monument. The area, the ruin, and its environment have been discussed by Caywood and Spicer (1935: 5-17). At its peak of growth the pueblo was composed of 110 rooms. Remains of three individual macaws were recovered.

Bones Recovered

V39 – Scarlet Macaw (*Ara macao*), Immature or older. Species identification on character of humerus.
Provenience: Surface.
Elements: Right humerus only.

V40 – Scarlet Macaw (*Ara macao*), Immature or older. Species identified on character of humerus.
Provenience: Room 27, Group V; trash fill in human burial.
Elements: Left humerus and right acetabulum, assigned by association in situ.

V958 – Scarlet Macaw (*Ara macao*), Newfledged (11-12 months old). Species identified on six diagnostic characters.
Provenience: Room 2, Group I, burial from beneath floor.
Elements: 62, head, neck, tongue, trachea, axial body, right wing and right leg.
Pathology: Right ulna roughened.
Remarks: Buried in the flesh (tracheal rings present).

Comments. Macaw V958 was buried in refuse upon which Room 2 of Group I was later built (Caywood and Spicer 1935: 95), so the burial date of V958 is before the "construction date" of Room 2, which appears to have been in the late 1300s. At least a timber believed to be from Room 2 has been dated 1386, with some rings missing; a timber from Group I, Room 12, has been dated 1366, also with some rings missing (Bannister, Hannah, and Robinson 1966: 12). Caywood informs me that the cultural material from rooms of Group I was the same, and that while excavating the group he considered it a unit-building, and still does. It seems logical, therefore, to consider that the macaw burial was made before 1366.

An analysis of pottery types and temporal sequences based on Colton and Hargrave (1937) also throws some light on the time the macaw was buried. The earliest ceramic series at Tuzigoot dates from about 1125 to 1200; a later series extends from about 1300 to 1400, the time of the late building surge (Caywood and Spicer 1935). These two temporal sequences also occurred at Montezuma Castle. The intervening 100 years, however, is not a period of abandonment as it might seem, because indigenous pottery types still occurred in abundance; but trade in vessels with painted decoration seemed to have ceased until about 1300. It seems highly improbable that if trade in pottery vessels was disrupted for an appreciable time, trade in macaws would not have been disrupted also. The most likely time that the macaw (V958) was buried would be during the period from 1125 to 1200.

MOGOLLON CULTURE AREA

Radiating at irregular distances from Alma, New Mexico, is a great expanse of territory in which are several areas with certain cultural traits in common, although each area is different in some respects from the others. The ceramic cultural traits are (1) applied decoration–paint (various colors), (2) surface color vessels–normally shades of brown or red, sometimes white, (3) surface finish, utility vessels–frequently corrugated, (4) firing atmosphere–oxidizing, and (5) method of thinning vessel walls–hammered (derived from paddle and anvil?).

Kinishba Pueblo

Kinishba Pueblo is about 8.5 miles southwest of Whiteriver, the headquarters of Fort Apache Indian Reservation, and 4 miles west of Fort Apache, on State Highway 73, in Gila County, Arizona. In the 1930s Kinishba was excavated under the direction of Byron Cummings of the University of Arizona, who published a popular account of the excavation and the material culture (Cummings 1940).

Remains of four individual macaws were recovered.

Bones Recovered

D813 – Scarlet Macaw (*Ara macao*), Newfledged (11-12 months old). Identified to species on premaxilla, basitemporal plate, quadrate, basihyal-basibranchial, manubrium, humerus, and tarsometatarsus.
Provenience: No provenience given.
Elements: 93, head, neck, tail, tongue and trachea, axial body, wings and legs.
Pathology: Carina of sternum crushed; left manus, digit 2, phalanx 1, tumorous; phalanx 3 fractured and tumorous.
Remarks: Macaw buried in the flesh (trachea present).

D814 – Scarlet Macaw (*Ara macao*), Newfledged (11-12 months old). Identified on species on six diagnostic characters.
Provenience: Room 65.
Elements: 92.
Pathology: Distal ends of tibiotarsi and tarsometatarsi eaten away by disease.
Remarks: Buried in the flesh (tracheal rings present).

D815 – Scarlet Macaw (*Ara macao*), Newfledged (11-12 months old). Identified to species on premaxilla, basitemporal plate, quadrate, manubrium, humerus, and tarsometatarsus.
Provenience: No provenience given.
Elements: 163, practically complete skeleton.
Pathology: Present.
Remarks: Buried in the flesh, tracheal rings present.

D830 – Scarlet Macaw (*Ara macao*), Adolescent (1-3 years old). Identified to species on basitemporal plate.
Provenience: No. 25576.
Elements: 2, premaxilla and skull.
Remarks: Probably bones remaining from a worn-out stuffed macaw.

Comments. In the absence of the provenience data, accurate correlative associations of macaw burials to material culture cannot now be determined for each of these four individual macaws; nor, by the same token, can temporal associations be made. However, 59 beams of reliable dates from Kinishba have given a time range from 1254 to 1299 (Smiley 1951).

Grasshopper Ruin, Arizona P:14:1

Grasshopper, a 14th–century ruin, is located about 10 miles west of Cibecue, Arizona, on Salt River Draw, just under the Mogollon Rim. There are more than 500 rooms, a sampling of which is being excavated by the University of Arizona. Geographic location and cultural parallels with Kinishba place Grasshopper Ruin in the Mogollon Culture Area.

Thompson and Longacre (1966: 255) report that "four bird burials were found just beneath the floor of the Great Kiva and at least one of these was a macaw burial." More recently Olsen (1967: 57) writes that "several macaw skeletons were recovered in 1966." These reports entitle Grasshopper Ruin to a place herein, although apparently no species identification has been made.

Beam specimens from the Great Kiva have been dated with a range extending from 1205, with some rings missing, to 1347, also with some rings missing (Bannister, Gell, and Hannah 1966: 52). There are no available data to determine a time relation of the kiva dates to the macaw burial from the kiva.

The Point of Pines Area

The Point of Pines Area is situated 60 miles east of Globe, Arizona, and lies at the abrupt transition from the pines of Nantack Ridge to the grasslands of Circle Prairie stretching north ten miles to Black River (Woodbury 1961: 1-3). Extensive excavations in many sites were carried out in this area from 1946 through 1960 by the University of Arizona Archaeological Field School (Robinson and Sprague 1965).

The two largest sites excavated, Turkey Creek (Arizona W:10:78) and Point of Pines Ruin (Arizona W:10:50), produced a total of 39 macaws.

Turkey Creek, Arizona W:10:78

Turkey Creek Pueblo, occupied between 1100 and 1200, is 3 miles northwest of Point of Pines, Arizona. It was excavated by the University of Arizona Archaeological Field School during 1958, 1959, and 1960 under the direction of Emil W. Haury and Raymond H. Thompson. Financial aid for the excavation was provided by the National Science Foundation. The large, single-story pueblo included about 325 contiguous domiciliary and storage rooms, two plazas, two small kivas, and a Great Kiva. A small contemporaneous structure was located on an adjacent knoll (Johnson 1965: iii, 60).

From Turkey Creek have come bones of 12 individual macaws, 7 herein identified as the Scarlet Macaw and 5 only to genus, *Ara,* because no species characters were present on the elements recovered.

Bones Recovered

D2206 – Macaw (*Ara* sp.). No species or age characters present.
Provenience: 329, trench mound. Small burial above 240.
Elements: Left coracoid.

D2207 – Macaw (*Ara* sp.). Newfledged (11-12 months). No species characters present.
Provenience: 300 north midden.
Elements: Sternum, right tibiotarsus.

D2211 – Macaw (*Ara* sp.). Immature (4-11 months). No species characters intact.
Provenience: 319, Trash Mound 2, Broadside 2.
Elements: Left humerus and right carpometacarpus.

D2215 – Macaw (*Ara* sp.), Newfledged (11-12 months). No species characters present.
Provenience: Broadside 6 and 3.
Elements: Right tibiotarsus.

D2216 – Macaw (*Ara* sp.), Immature (4-11 months). No species characters present.
Provenience: Broadside 6 and 3.
Elements: Right tibiotarsus.
Pathology: Shaft of tibiotarsus abnormally curved, apparently resulting from an early greenstick fracture.

D2208 – Scarlet Macaw (*Ara macao*), Adolescent (1-3 years). Species identification based on basitemporal plate.
Provenience: 302,Trash Mound 1, Broadside 1.
Elements: Cranium, right humerus, right and left ulnae, and left femur.
Pathology: Diseased area around aperture of the right Eustachean tube of cranium.

D2209 – Scarlet Macaw (*Ara macao*), Newfledged (11-12 months). Species identification based on basitemporal plate.
Provenience: 310, Trash Mound 1, Broadside 1.
Elements: Cranium, mandible, left radius, left femur, and right tibiotarsus.
Remarks: Radius, rodent-gnawed.

D2210 – Scarlet Macaw (*Ara macao*), Adolescent (1-3 years). Species identification based on three characters: basitemporal plate, humerus, and manubrium.
Provenience: 310, Trash Mound 1, Broadside 1.
Elements: 8, representing head, axial body, and left wing.
Pathology: Right and left quadrates fused to cranium; deposits of calcium on articular surfaces of mandible; left ulna fused to humerus.

D2212 – Scarlet Macaw (*Ara macao*), Newfledged (11-12 months). Species identification based on two characters: basitemporal plate and manubrium.
Provenience: 314, Trash Mound 5, Broadside 4.
Elements: Cranium and sternum.

D2213 – Scarlet Macaw (*Ara macao*), Adolescent (1-3 years). Species identification based on three characters: premaxilla, basitemporal plate, and humerus.
Provenience: 328, Trash Mound 6, Broadside 5, Turkey Burial 5.
Elements: 8, representing head, trachea, and right wing.
Pathology: Proximal head of right humerus crushed and healed; distal end of humerus and proximal head of ulna display adjacent abcesses; ulna greatly roughened.
Remarks: Buried in the flesh (tracheal rings present).

D2214 – Scarlet Macaw (*Ara macao*), Newfledged (11-12 months). Species identification on basis of character of humerus.
Provenience: 313, Trash Mound 6, Broadside 5, Turkey Burials 1 and 2.
Elements: 4, representing head and both wings.
Pathology: Ulna slightly roughened.
Remarks: Age assigned on degree of development of tendinal groove of ulna.

D2217 – Scarlet Macaw (*Ara macao*), Newfledged (11-12 months). Species identification

based on five characters: premaxilla, basitemporal plate, quadrate, manubrium, and humerus.
Elements: 18, representing head, neck, axial body, and wings.

Point of Pines Ruin, Arizona W:10:50

The occupation of Arizona W:10:50, a thousand-foot-long pueblo of 600 to 800 rooms, extended from 1200 to about 1450, according to Robinson (1958:13). No statement was made as to whether this was ceramic or tree-ring dating. However, Haury (1958: 4) gives tree-ring dates of 1262-93, with a clustering of 80% in 1280-85.

The 27 individual macaws represented by bones recovered from Arizona W:10:50 can be assigned to the late 14th century according to E. W. Haury and R. H. Thompson. Fifteen of these were identified as the Scarlet Macaw (*Ara macao*); bones of 12 were lacking in species characters.

Bones Recovered

D2183 – Macaw (*Ara* sp.), Breeding (4-? years). No species characters.
Provenience: Room 31, Subfloor 1, with D2184.
Elements: 3 fragments, mandible.
Remarks: Mandible very heavy, without perforations characteristic of younger birds (Fig. 19).

D2185 – Macaw (*Ara* sp.), Immature (4-11 months). No species characters.
Provenience: Room 31, Subfloor 1, above Floor 2.
Elements: Right tibiotarsus.

D2188 – Macaw (*Ara* sp.), Newfledged (11-12 months). No species characters.
Provenience: Room 62, Level 3.
Elements: 18, representing head, neck, axial body, one wing, and legs.

D2191 – Macaw (*Ara* sp.), Adolescent (1-3 years). No species characters.
Provenience: Room 62, floor, No. 133.
Elements: 24, representing head, neck, axial body, wings and right leg.
Pathology: Roughened ulnae.

D2192 – Macaw (*Ara* sp.), Newfledged (11-12 months). No species characters.
Provenience: Room 62, floor.
Elements: 4, representing right wing.
Pathology: Ulna roughened.

D2193 – Macaw (*Ara* sp.), Newfledged (11-12 months). No species characters.
Provenience: Room 62, Level 3.
Elements: Right ulna, left radius, and right tibiotarsus.
Pathology: Ulna roughened and proximal head diseased.

D2195 – Macaw (*Ara* sp.), Immature (4-11 months). No species characters.
Provenience: Room 62, Level 4, NE Quad.
Elements: Fragments of sternum and radius.
Remarks: Aged, on thinness of bone and faintness of intermuscular lines of sternum.

D2197 – Macaw (*Ara* sp.). No age or species characters present.
Provenience: Great Kiva, NW quarter, subfloor, general.
Elements: Right ulna.
Pathology: Ulna roughened.

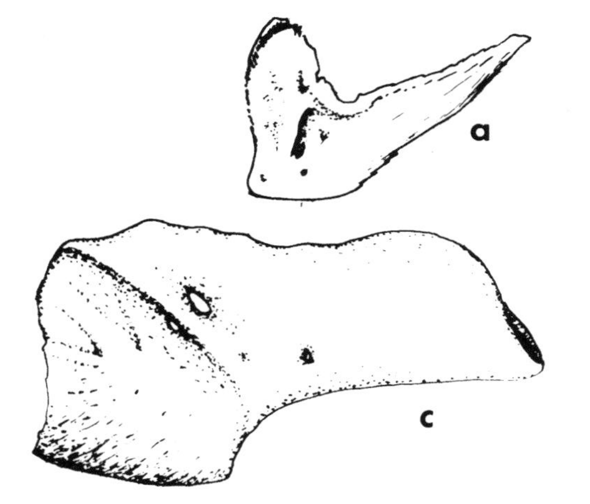

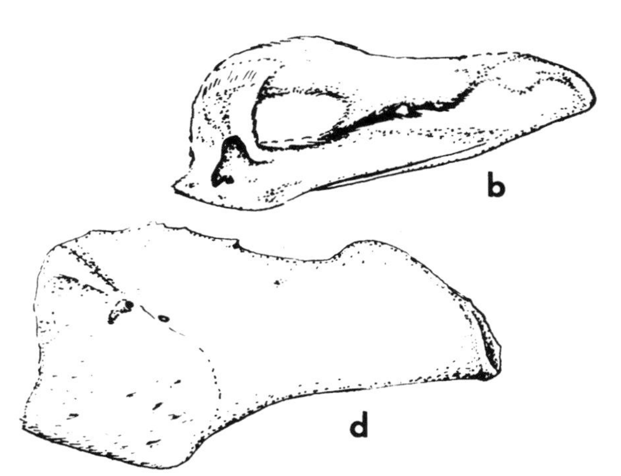

Fig. 19. Development of the mandible. *a, Ara chloroptera* (H2451) mandible, Juvenile (7 weeks); *b, Ara chloroptera* (H2449) mandible, Juvenile (8 weeks); *c, Ara macao* (A0534) mandible, Immature (4-11 months); *d, Ara macao* (H2159) mandible, Newfledged (11 to 12 months). (Scale: X 1)

D2199 – Macaw (*Ara* sp.). No age or species characters present.
Provenience: No. 15.
Elements: Damaged left humerus.

D2201 – Macaw (*Ara* sp.), Newfledged (11-12 months). No species characters.
Provenience: Loose, No. 158.
Elements: Left tibiotarsus.

D2202 – Macaw (*Ara* sp.), Newfledged (11-12 months). No species characters.
Provenience: Loose, No. 94.
Elements: Left tibiotarsus.

D2204 – Macaw (*Ara* sp.), Newfledged (11-12 months). No species characters.
Provenience: Loose, No. 130.
Elements: Damaged cranium and mandible.

D2179 – Scarlet Macaw (*Ara macao*). Species identification based on character of humerus. No age characters present.
Provenience: Room 3, No. 17.
Elements: Left humerus.
Pathology: Attachment of anterior articular ligament at distal end of humerus surrounded by accretions.

D2180 – Scarlet Macaw (*Ara macao*), Immature (4-11 months). Species identification based on basitemporal plate of cranium, age on suborbital arch.
Provenience: Room 4, No. 3.
Elements: Cranium.

D2181 – Scarlet Macaw (*Ara macao*), Newfledged (11-12 months). Species identification based on seven characters: premaxilla, basitemporal plate, quadrate, humerus, basihyal-basibranchial, manubrium, and tarsometatarsus.
Provenience: Room 11, floor.
Elements: 50, representing entire bird.
Pathology: Many bones display accretions, especially at attachments of tendons.
Remarks: Buried in the flesh (tracheal rings present).

D2182 – Scarlet Macaw (*Ara macao*), Newfledged (11-12 months). Species identification based on character of the premaxilla; age based on bone density.
Provenience: Room 11, floor.
Elements: Premaxilla.

D2184 – Scarlet Macaw (*Ara macao*), Newfledged (11-12 months). Species identification based on character of tarsometatarsus, age on characters of cranium.
Provenience: Room 31, Subfloor 1.
Elements: 18, representing head, trachea, axial body, wings, and right foot.
Pathology: Ulnae roughened.
Remarks: Buried in the flesh (tracheal rings present).

D2186 – Scarlet Macaw (*Ara macao*), Immature (4-11 months). Species identification on six characters: premaxilla, basitemporal plate, quadrate, basihyal-basibranchial, humerus, and manubrium.
Provenience: Room 31, Level 15 cm. above Floor 2.
Elements: 71, representing entire bird.
Pathology: Palatine fused to pterygoid; proximal and distal heads of humeri diseased; left ilium broken and healed at right angle; right tibiotarsus twisted proximally and fractured distally, healing without rejoining distal end; right manus, digit 2, phalanx 1 damaged.
Remarks: Buried in the flesh (tracheal rings present).

D2187 – Scarlet Macaw (*Ara macao*), Newfledged (11-12 months). Species identification based on six characters: premaxilla, basitemporal plate, quadrate, basihyal-basibranchial, manubrium, and tarsometatarsus.
Provenience: Room 62, Level 3, Roof 1.
Elements: 49, representing head, neck, tail, tongue, trachea, axial body, left wing, and legs.
Pathology: Premaxilla fusing to cranium; peck injury to premaxilla (Fig. 20*a*).
Remarks: Buried in the flesh (tracheal rings present).

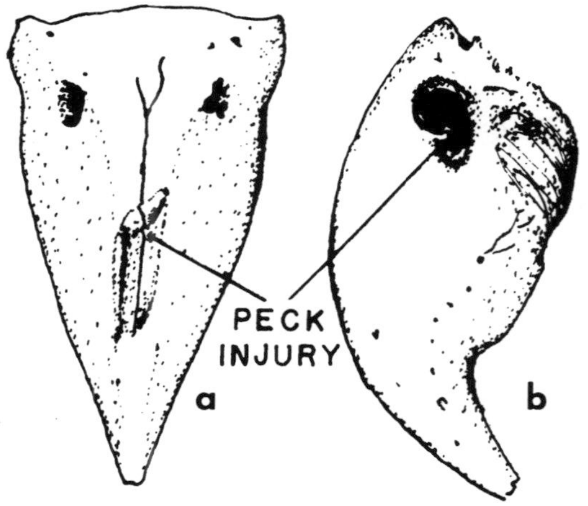

Fig. 20. Peck injury; premaxillae displaying destruction of bone caused by being pecked by other macaws. *a, Ara macao* (#5226), anterior view, depression along culmen; *b, Ara macao* (#5234), lateral view, external naris enlarged by peck. (Scale: X 1)

D2189 – Scarlet Macaw (*Ara macao*), Breeding (4-? years). Species identification based on character of tarsometatarsus; age based on characters of tibiotarsus.
Provenience: Room 62, Level 3.
Elements: Right tibiotarsus and right tarsometatarsus.
Pathology: Distal end of tibiotarsus diseased.

D2190 – Scarlet Macaw (*Ara macao*), Newfledged (11-12 months). Species identification based on basitemporal plate, cranium, humerus, manubrium, and tarsometatarsus.
Provenience: Room 62, Level 3.
Elements: 112, representing entire bird.
Pathology: Ulnae greatly roughened. Accretions on distal end, right humerus; coracoidal head, right scapula; humeral head, right coracoid; right opisthotic process of cranium; and right quadrate.
Remarks: Buried in the flesh (tracheal rings present).

D2194 – Scarlet Macaw (*Ara macao*), Newfledged (11-12 months). Species identification based on five characters: premaxilla, basitemporal plate, quadrate, manubrium, and tarsometatarsus.
Provenience: Room 62, Level 3.
Elements: 74, representing entire bird.
Pathology: Roughened ulna.
Remarks: Buried in the flesh (tracheal rings present).

D2196 – Scarlet Macaw (*Ara macao*), Newfledged (11-12 months). Species identification based on four characters: manubrium, humerus, basihyal-basibranchial, and tarsometatarsus.
Provenience: Room 62, Level 4, NE Quad.
Elements: 122, representing entire bird.
Pathology: Ulnae slightly roughened; right and left fibulae fusing to right and left tibiotarsi; carinal margin of sternum crushed and healed.
Remarks: Buried in the flesh (tracheal rings present).

D2198 – Scarlet Macaw (*Ara macao*), Newfledged (11-12 months). Species identification on character of humerus.
Provenience: No. 10.
Elements: Premaxilla, one tracheal ring, left humerus, right tibiotarsus.
Remarks: Buried in the flesh (tracheal rings present).

D2200 – Scarlet Macaw (*Ara macao*), Newfledged (11-12 months). Species identification based on character of tarsometatarsus.
Provenience: Loose.
Elements: right femur, right tibiotarsus, and right tarsometatarsus.

D2203 – Scarlet Macaw (*Ara macao*), Newfledged (11-12 months). Species identification based on character of premaxilla.
Provenience: Loose, No. 123A.
Elements: Cranium and premaxilla.

D2205 – Scarlet Macaw (*Ara macao*), Newfledged (11-12 months). Species identification based on characters of basitemporal plate and humerus.
Provenience: Loose.
Elements: Cranium, right humerus, and left tibiotarsus.
Remarks: Tibiotarsus, rodent-gnawed.

Reeve Ruin

The Reeve Ruin is located by the San Pedro River about 3.5 miles southeast of Redington, Pima County, and about 40 miles down river (north) from Benson, Cochise County, Arizona. The excavation of this site was a project of the Amerind Foundation, under the directorship of Charles C. DiPeso. From published descriptions and comments (DiPeso 1958), it is apparent that the tradition of the indigenous ceramics is Mogollon derived. Two bones of one individual macaw were found.

Bones Recovered

D2220 – Macaw (*Ara* sp.), Newfledged (11-12 months old). No species character present.
Provenience: In Plaza 2, in front of room 6, building phase 2, of House Block 2 (Di Peso 1958: 43, Fig. 7); bag 10.
Elements: Distal halves of right ulna and right radius.
Pathology: Ulna roughened.
Remarks: Age assigned on the stage of development of the tendinal groove of the ulna. Bones appear ordinary in all other respects. The previously published identification is an error (DiPeso 1958: 116)! These bones were in a crushed Belford Plain jar.

Comments. That these bones were found buried in dirt in a jar should not be considered proof that macaws were used as food by Southwestern Indians. I have not seen a single humanly altered macaw bone from an archaeological site. Dates for the building and occupation period of Reeve Ruin have not been determined from tree-rings, nor is there a range of ceramic dates available for intrusive trade vessels. DiPeso has informed me that the time of occurrence of this macaw, therefore, cannot be established, but is considered to be very late.

Freeman Ranch Site

The Freeman Ranch Site is about 5 miles north of Cliff, New Mexico, on a low mesa above Duck Creek. Here was found the burial of an elderly woman, extended on the back, and with macaw bones located above the left shoulder extending to the top of the head. Elsewhere bones of another macaw were found, but in such extremely poor condition that they could not be saved. No pottery was associated with either of these macaw burials.

Bones Recovered

D2221 – Scarlet Macaw (*Ara macao*), Newfledged (11-12 months old). Species identification based on premaxilla, basitemporal plate, quadrate, manubrium, and humerus.
Provenience: 291FR-6; Below cremation area on southwest slope below FR-1A.
Elements: 19, representing head, tail, axial body, left wing, and legs.
Pathology: Left ulna roughened.

A definite relationship exists between the macaw buried in one of the rooms of the Freeman Site and the Salado ceramic complex, since the building is of the Salado culture. Moreover, this Gila Polychrome occupation is superimposed on an abandoned Mimbres site, so this particular macaw is post-Mimbres, or post 1150. Nesbitt (1931: 100) says "examples of this superposition are plentiful in the Mimbres Valley and more instances might be cited."

MIMBRES CULTURE AREA

The Mimbreños occupied the drainage of the Mimbres River in Southwestern New Mexico. Their ceramic cultural indicators are: (1) applied decoration–paint color (black), geometric and zoomorphic designs; (2) surface color–normally shade of gray or white; (3) surface finish utility vessels–frequently plain with corrugated necks, braided handles; (4) firing atmosphere–reducing; and (5) method of thinning vessel walls–hammering.

Cameron Creek Village

After a period of apparent abandonment beginning about 950, Cameron Creek Village was occupied by a group of Mimbreños living in surface masonry structures and using the characteristic Mimbres Black-on-white pottery. Macaw remains were recovered from Surface Room 60 and from the dump stratum associated with this occupation.

No. 1 – Macaw (*Ara macao*). Identified by Wetmore.
Provenience: With a human burial in the deep black soil fill of Room 60.
Elements: Not available for study.
Remarks: This macaw was found over the left arm of an adult male human skeleton (Bradfield 1929: 11).

Comments. In addition to the macaw, eight pottery vessels were found in Room 60. These vessels, according to Bradfield's descriptions and illustrations, are Mimbres Black-on-white. Haury (1936: 116, 12a) places the beginning date for Mimbres Black-on-white as soon after 1000 and its terminal date at about 1150, at which time he believes the Mimbreños to have abandoned the valley. Thus, it is likely that Macaw No. 1 was buried between 1000 and 1150.

No. 2 – Macaw (*Ara* sp.)
Provenience: "Recovered from the black soil fill in the lower east side of the west group" (Bradfield 1929: 11).
Elements: Not available for study.

Discussion. The dump stratum from which Macaw No. 2 was recovered pertains to Construction Stage VI, dated by ceramics at about 1000 to 1150.

Galaz Site, New Mexico

The Galaz Site of southwestern New Mexico was excavated by Albert Jenks of the Department of Anthropology, University of Minnesota, from 1929 to 1931. Most of the macaw remains were recovered from Kiva 73. Sherds from beneath the floor of this kiva are predominantly Mimbres Black-on-white, placing the use of this kiva toward the latter part of Mimbres Culture. A date for the deposition of the macaw remains probably would be between 1100 and 1150, based on the predominant associated ceramic type, classic Mimbres Black-on-white, and on the abandonment of the Mimbres Valley probably about 1150.

From the occurrence of Scarlet Macaws in Chaco Canyon, it is evident that this single occurrence of the Military Macaw is contemporaneous with the presence of the Scarlet Macaw here and elsewhere.

Bones Recovered

D2268 – Military Macaw (*Ara militaris*), Newfledged (11-12 months).

Provenience: Under green stone in floor of Kiva 73.
Elements: About 25, representing the entire carcass.
Pathology: Roughened ulna.

D2269 – Scarlet Macaw (*Ara macao*), Newfledged (11-12 months).
Provenience: Kiva 73.
Elements: About 13, representing head, neck, axial body, right wing, and right leg.
Pathology: Roughened ulna, 2 caudal vertebrae fused.

D2270 – Scarlet Macaw (*Ara macao*), Newfledged (11-12 months).
Provenience: Kiva 73.
Elements: About 7, representing right wing, and right and left legs.
Pathology: Two spurs anterior and ventral to condyle at foramen magnum.

D2271 – Scarlet Macaw (*Ara macao*), Newfledged (11-12 months).
Provenience: Area north of Room 35.
Elements: About 11, representing head, axial body, right wing and leg.
Pathology: Roughened ulna.

HOHOKAM CULTURE AREA

The Hohokam Culture Area centers along the Gila and Salt rivers just above their confluence. Closely connected culturally are the neighboring regions surrounding Tucson and comprising the present Papago Reservation. The ceramic cultural indicators are (1) applied decoration–red paint, (2) surface color–normally "buff," (3) surface finish utility vessels–noncorrugated, (4) firing atmosphere–oxidizing, and (5) method of thinning vessel walls–paddle and anvil.

Gatlin Site, Arizona Z:2:1

Three miles north of Gila Bend, Arizona, and half a mile south of the Gila River, the Gatlin Site was excavated between November 1958 and February 1959 by the Arizona State Museum for the National Park Service under the direction of William Wasley. The platform mound, displaying six major stages of development, was an earth-cored construction with a flat top, rounded corners, and sloping sides faced with caliche and adobe plaster. The entire series of alterations is placed within the Sacaton Phase of the Hohokam Culture, about 900 or 950 to 1100 or 1150. It is considered to be a northern manifestation of the Mesoamerican pyramid-temple complex (Wasley 1960: 244).

Bones Recovered

D2005 – Scarlet Macaw (*Ara macao*), Newfledged (11-12 months). Species identification on manubrium.
Provenience: Trench 1, just outside cobble facing of mound.
Elements: Poorly preserved fragments of 30, representing head, neck, axial body, wings, and legs.

Comments. The Sacaton Phase, from 950-1150 in the Gila Bend area, included a premound occupation after 950, the hundred-year span of platform construction, and a period of abandonment preceding 1150. Wasley informs me that since the macaw was found just outside the cobble facing of the final stage of platform construction, it most probably was deposited during or about 1150.

RIO GRANDE PUEBLO AREA

A number of macaws have been found at several protohistoric and historic Rio Grande Pueblos.

Picuris Pueblo

Picuris Pueblo (San Lorenzo) lies on the north side of Rio Pueblo, a permanent east-west flowing stream rising in the Sangre de Cristo Mountains of northern New Mexico. The nearest modern community is Peñasco, New Mexico. With Taos and Pecos, Picuris formed the eastern boundary of the northern Rio Grande Pueblo Culture. Picuris appears to have been a center of trade, carrying mountain products to the lowland pueblos in the south.

Picuris Pueblo is reported to have had a population of 3,000 people in 1600, occupying two large house blocks, each four or five stories high. The population in the 1960s has shrunk to about 100 persons, who have cooperated with the Fort Burgwin Research Center in the excavation of the pueblo. The work has been carried on under the direction of Herbert W. Dick, of Adams State College of Colorado, with additional financial assistance from the National Science Foundation and the National Park Service.

The Pueblo was most actively occupied beginning about 1275 and has been continuously occupied into the 1960s. A macaw bone was recovered from a mixed deposit yielding ceramic dates of 1300 to 1700.

Bones Recovered

PP424 – *Ara* sp. (Macaw).
Provenience: TA III, Area II, Feature 12, Level 2.
Element: Left ulna.
Pathology: Ulna roughened.

Pecos Pueblo

Pecos Pueblo, San Miguel County, New Mexico, was located on the top of a steep-sided sandstone mesa bordering the eastern bank of Arroyo del Pueblo, which flows into the Pecos River.

Its occupation began about 1200 with the abandonment of Forked Lightning Ruin across the arroyo in favor of this more easily protected site. The pueblo, occupied continuously until its abandonment in 1838, was one of the most important settlements in the Southwest in late prehistoric times, since it controlled the best pass between the Pueblos and the Plains. This trade center appears to have been most active between 1300 and 1700 (Kidder 1931: 3-8).

During the excavations conducted between 1915 and 1925, by A. V. Kidder under the auspices of the Robert S. Peabody Foundation for Archaeology, Phillips Academy, Andover, Massachusetts, two macaw skeletons were recovered from the same, presumably human, grave.

A date of deposition for these bones, based on the activity peak of the settlement would be between 1300 and 1600 (Kidder 1924: 4).

Bones Recovered

291284A – *Ara macao* (Scarlet Macaw), Newfledged (11-12 months). Species identification on premaxilla, basitemporal plate, manubrium, humerus.
Provenience: "Two skeletons from same grave (mixed in earth)."
Elements: 24, representing whole bird.
Pathology: Many fractured bones which have healed.
Remarks: Buried in the flesh (tracheal rings present).

291284B – *Ara macao* (Scarlet Macaw), Newfledged (11-12 months). Species identification on premaxilla, basitemporal plate, quadrate, manubrium, humerus, tarsometatarsus.
Provenience: "Two skeletons from same grave (mixed in earth)."
Elements: 26, representing the whole bird.
Remarks: Buried in the flesh (tracheal rings present).

García Site, Pojoaque Pueblo

In 1953 the University of New Mexico Archaeological Field School, under the direction of Florence Ellis, excavated a portion of the García Site at Pojoaque Pueblo, New Mexico. The site consists of a dump behind the church and perhaps 35 rooms of the ruin left after the remainder of the area was leveled for modern farming activities.

Although portions of Pojoaque had been inhabited from Pueblo II times, the García Site was not occupied until about 1400. From the fill of Room 31, in the northwest corner of the remaining room block, the upper half of the beak of a Scarlet Macaw was recovered. Ceramics from Room 31 included Biscuit A, from trash below the floor, and Sankawi Black-on-cream, Polished Red, and Polished Black from the room fill. Dates from these ceramics place the latest use of the room at about 1600. Ellis places the disposition of the macaw bone between 1400 and 1600.

Bone Recovered

D766 – Scarlet Macaw (*Ara macao*). Identified to species on premaxilla.
Provenience: Room 31, fill.
Elements: Premaxilla.

Comments. Although Oñate's people came into the area in 1598, this portion of the Pojoaque settlement is considered to be pre-Spanish.

Gran Quivira

Gran Quivira, the Pueblo de las Humanas, was in early historic times one of the largest pueblos in central New Mexico and marked the southeastern limit of the Pueblo Area. It was apparently abandoned in 1672 (Vivian 1964: 9, 10 & 3).

Archaeological excavations were carried on by E. L. Hewett in 1923 and 1925, by Wesley Bradfield in 1924, and by Gordan Vivian in 1951. No mention of macaw remains from any of these excavations has been found (Vivian 1964: 5). Excavations conducted by Alden Hayes for the National Park Service from 1965 through 1967 have resulted in recovery of the skeleton of the Scarlet Macaw from Mound 7.

Bones Recovered

Q4 – *Ara macao* (Scarlet Macaw), Newfledged (11-12 months). Species identification was made on premaxilla, basitemporal plate, basihyal-basibranchial, quadrate,

manubrium, humerus, and tarsometatarsus.
Provenience: Mound 7, Room 182, under floor (F S9701).
Elements: 69, representing the entire bird.
Pathology: Ulna roughened; accretion on distal head of right tibiotarsus; destruction of bone along the premaxillary hinge line of the cranium.
Remarks: Buried in the flesh (tracheal rings present).

Comments. Room 182 was part of a structure associated with a kiva built in 1416. Hayes informs me that there was no evidence of pit outlines indicating connection with the occupations of Room 182, and that the depth is much greater than is usual for under-floor burials of macaws. It appears, therefore, most likely that the macaw was deposited in trash fill before the construction of Room 182, or prior to 1416.

TABLE 9

Summary of Macaw Archaeological Specimens Analyzed in This Study

Area and Site	Species				Pathology	Tracheal Rings	Age Stages				
	Ara macao	*Ara militaris*	*Ara* sp.	Total			3 Immature	4 New-fledged	5 Adolescent	6 Breeding Age	7 Aged
Anasazi Area											
Pueblo Bonito	30		1	31	19	4	1	21	7		1
Pueblo del Arroyo	3			3	2	2		3			
Kin Kletso	1			1					1		
Houck K	1			1				1			
Kiet Siel			2	2	1			2			
Sinagua Area											
Wupatki	22		19	41	12	8	4	15		1	
Nalakihu			1	1					1		
Ridge Ruin	3		1	4	1	1	2	2			
Winona Village			1	1							
Pollock Site	1			1	1			1			
Montezuma Castle	1			1	1			1			
Jackson Homestead	1			1					1		
Tuzigoot	3			3	1	1	2	1			
Mogollon Area*											
Kinishba	4			4	3	3		3	1		
Turkey Creek	7		5	12	5	1	2	6	3		
Point of Pines Ruin	15		12	27	13	8	4	17	1	2	
Reeve Ruin			1	1	1			1			
Freeman Ranch	1			1	1			1			
Mimbres Area†											
Galaz Site	3	1		4	4			4			
Hohokam Area											
Gatlin Site	1			1				1			
Rio Grande Area											
Picuris			1	1	1						
Pecos	2			2	1	2		2			
García Site	1			1							
Gran Quivira	1			1	1	1		1			
Total	101	1	44	145	68	31	15	83	15	3	1

* Seven macaws from Grasshopper (Arizona P:14:1) were not analyzed.

† Two macaws (one *A. macao* and one *A.* sp.) from Cameron Creek Village were not available for analysis.

4. SUMMARY

In the first part of the study seven osteological age classes are established for the Mexican macaws: Nestling (hatching to 6 weeks), Juvenile (7 weeks to 4 months), Immature (4 to 11 months), Newfledged (11 to 12 months), Breeding Age (over 4 years), and Aged (very old). Species characters to distinguish *Ara militaris* from *Ara macao* are defined on the basis of the following elements: premaxilla, cranium, quadrate, basihyal-basibranchial, sternum, humerus, and tarsometatarsus. Residual characters are assigned to differences of geographic areas, sex, effects of captivity, pathology, and individual variation.

The second part of the study is a comparative examination and review of all available skeletal material from southwestern archaeological sites north of the international border. A total of 145 macaws was examined. Of these, 101 were identified as *Ara macao* (Scarlet Macaw) and one as *Ara militaris* (Military Macaw); 44 were too fragmentary to identify as to species (Table 9).

Of the 145 individuals, 68 (or 47%) displayed pathological bones (Table 10). The pathological conditions reflected normal accidents or were attributable to dietary deficiencies. There is no pattern suggestive of deliberate mutilation, such as breaking a wing to inhibit flight. Tracheal rings, indicative of burial in the flesh, were recovered from only 31 individuals (or about 21% of the collection). In all probability they were present in others, but because of their fragile nature were not recovered.

Analysis of these and other data indicate that only one Military Macaw (*Ara militaris*) was present (at the Galaz Site) in the collections examined and that all other prehistoric macaws identifiable to species were Scarlet Macaws (*Ara macao*), a tropical species.

The age classification of the 117 specimens that were complete enough to age is given in Table 10. The study of age characteristics indicated only three birds of breeding age (at Wupatki and Point of Pines), no nestlings or juveniles, and only one individual of advanced age (at Pueblo Bonito). The four older birds were generally isolated examples found as inhumations or associated with human burials, and suggest personal pets. The greatest number of specimens is assigned to the age range of slightly less than eleven months old to slightly over one year of age at time of death (Table 9).

Since macaws do not breed before their fourth to fifth year, and no nestlings or juveniles were found, all birds must have been obtained from trade to the south. Of the young fledged birds occurring with greatest frequency, those in the lower limits of the age group, less than eleven months old, occurred in higher frequencies in the southern portions of the Southwest while those twelve months old or older generally occurred in more northerly areas.

As no breeding of these birds can be shown to have occurred in the American Southwest, the data suggest a well-organized trade in young macaws originating farther to the south, in tropical Mexico. This trade in young macaws may have passed through northern Mexican centers of commerce and use, with some macaws eventually reaching northern Arizona and New Mexico in prehistoric times.

Concentrations of these birds north of the border seems to correlate with known centers of population, namely, Chaco Canyon in New Mexico, Wupatki and the San Francisco Mountains, and the Point of Pines and Mogollon Rim country of Arizona, and in about that order. Secondary radiation from these centers may have occurred, but the numbers of macaws reaching the smaller villages drops appreciably. Trade in macaws was established in the Southwest about 1100, or shortly before, and appears to have declined after about 1375. A late and seemingly weak revival of this trade following the Conquest is indicated for the Rio Grande Valley in New Mexico (Fig. 21).

TABLE 10

Macaw Archaeological Specimens Classified According to Age Stage

Age Stage	Individuals	Percentage
1. Nestling (0-6 wk)	0	0.0
2. Juvenile (7 wk - 4 mo)	0	0.0
3. Immature (4-11 mo)	15	12.8
4. Newfledged (11-12 mo)	83	71.0
5. Adolescent (1-3 yr)	15	12.8
6. Breeding Age (4+ yr)	3	2.6
7. Aged	1	0.8
	117	100.0

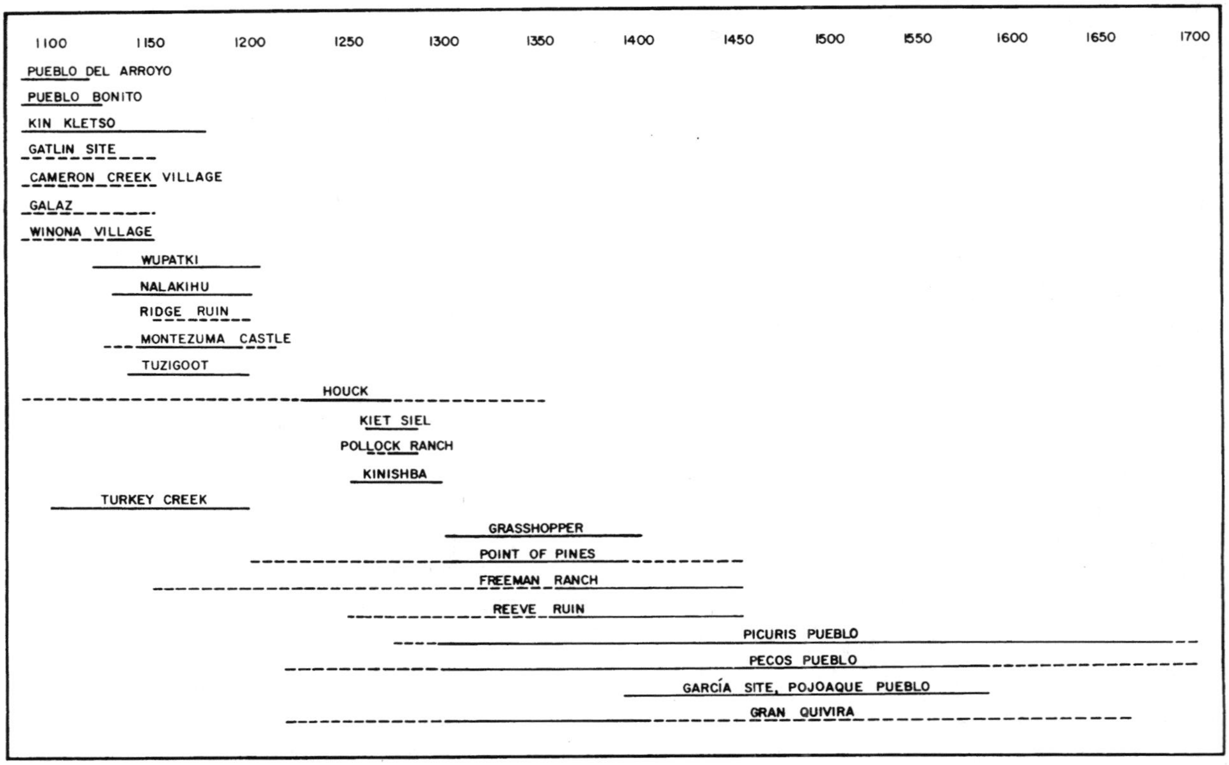

1100 1150 1200 1250 1300 1350 1400 1450 1500 1550 1600 1650 1700

PUEBLO DEL ARROYO
PUEBLO BONITO
KIN KLETSO
GATLIN SITE
CAMERON CREEK VILLAGE
GALAZ
WINONA VILLAGE
WUPATKI
NALAKIHU
RIDGE RUIN
MONTEZUMA CASTLE
TUZIGOOT
HOUCK
KIET SIEL
POLLOCK RANCH
KINISHBA
TURKEY CREEK
GRASSHOPPER
POINT OF PINES
FREEMAN RANCH
REEVE RUIN
PICURIS PUEBLO
PECOS PUEBLO
GARCÍA SITE, POJOAQUE PUEBLO
GRAN QUIVIRA

Fig. 21. Probable chronological distribution of macaw archaeological remains.

Appendixes

Appendix A

Confusion from Improper Use of Common Names of Mexican Macaws

The recording of field notes on bones of macaws, would, in general, be the same procedure as for most bones, as also would be the proper use of the upper and lower case letters for use in the published report. If the approved procedure were followed at all times, better understanding would be maintained. The proper use of scientific names, that is, Latinized names, is common knowledge to most students concerned with forms of natural history, but the correct use of English names is sometimes puzzling. "In general an upper case letter at the beginning of a name signifies a species of bird while a lower case letter, instead, refers to a group of birds" (Hargrave 1968).

For instance, common names of species are considered as proper nouns, as are the names of people. Therefore, in speaking of the "Scarlet Macaw" it is always written so; but if spoken of correctly and collectively, with reference to species, it could be written thus: "Scarlet Macaws are common." If reference is *not* to species but to a descriptive term (as color), it would be written thus: "green macaw" or "scarlet macaw." In a scientific report it would be written that : "Although a green macaw is native to Mexico the real Green Macaw (*Ara ambigua*) is from South of Mexico."

In 1890, six years before the beginning of the Hyde excavations at Pueblo Bonito, Bandelier (1890: 62) said of Mexico:

> A species of large green parrot inhabits the pine forests of the Sierra Madre as far north as latitude 30° [compare Van Rossem 1945: 101]. I have often seen it in the solitudes of the heart of this vast chain, early in the morning fluttering from tree-top to tree-top, and saluting the dawn of the day with its discordant cries and animated conversation.

Bandelier comments further, in a footnote, that the bird is called "Guacamayo or macaw, by the natives." He was wise, though, in indicating his doubt in the same footnote that the "large green parrot of the pine forests of the Sierra Madre was the green macaw (*Ara ambigus*) [sic] catalogued by Jesús Sánchez (in *Anales del Museo Nacional*, Vol. 1, No. 2, p. 94) as extant in the Southern Mexican States. . . ." Further, he says with qualification that he has "seen the *A. ambigus* [sic] in Oaxaca, and the green parrot in the heart of the Sierra Madre, west of the Cases Grandes (on the summit of the Puerto de San Diego)." It is obvious then that Bandelier himself was confused about the three "large green parrots," which today are known by size as: The Thick-billed Parrot (*Rhynchopsitta pachyrhyncha*), the Military Macaw (*Ara militaris*), and the Great Green Macaw (*Ara ambigua*) which today is not known to occur north of Nicaragua (De Schauensee 1964: 103; Peters 1937: 182). Moreover, not once did Bandelier write of any of these three green parrots as the Green Macaw and follow that "name" by the Latin name "*Ara militaris*," but, as stated, he did recognize the occurrence of a large green macaw near the United States border.

From the recovery of a tropical macaw skeleton from Precolumbian occupation horizons in the Chaco and contiguous areas, it is apparent that trade items were moved over distances of many hundreds of miles. If items were moved so far north from southern Mexico, items likewise could be moved from even further south, into Mexico; so that, with the increase of archaeological interest in biological remains from Mesoamerican ruins, bones of Guatemalan birds conceivably could be found in Mexico. It seems advisable at this time then to forestall misunderstandings in name, common or scientific, as, for instance, the correct common name for Bandelier's "*Ara ambigus*" now is *Ara ambigua* (Peters 1937: 182) a green macaw that already has confused the problem of common names in Mexico and even in the United States. Acceptance of de Schauensee's choice (1964: 103) of the name "Great Green Macaw" for the green macaw *Ara ambigua,* and the acceptance of the now favored common name of "Military Macaw" for the green macaw *Ara militaris* (Friedmann and others 1950: 124), would avoid further confusion about the common name of the Military Macaw (*Ara militaris*). Nevertheless, Bandelier's comments on the large green parrots of bordering Mexico would indicate that the large parrot most readily available to the Southwestern United States would be the nearest one, the one in Northern Mexico, the Military Macaw. Twenty years after publication of Bandelier's opinion, Pepper himself (1920: 195) said the same thing in relation to the parrot bones of Pueblo Bonito:

The *Ara militaris* or green macaw is found at the present time in certain parts of Mexico and there is strong evidence that it was at one time quite common in the northern part of Mexico and extended even to the southern parts of New Mexico and Arizona.

It should be noted here that no one has proved that any macaw has occurred as a native wild bird in the area now known as Arizona and New Mexico, nor is the Military Macaw listed in the *Check-List of North American Birds* (American Ornithologists' Union 1957: 267).

Appendix B

Necessity for Preservation of all Faunal Remains

In *A Plea for More Careful Preservation of All Biological Material From Prehistoric Sites* (Hargrave 1938), I point out the need for saving all bone material found in archaeological situations in order to determine how uses of biological materials have influenced local man for good or for bad. More importantly, however, we need to know how these lifeforms may have reflected the presence of intangible factors of nature, such as temperature and precipitation, since data on these variables contribute to our knowledge of past ecological conditions. These interrelated data, tangible and intangible, also help us to understand the vagaries of the economics of a given people during a given period of time. We can identify bones to species; we can frequently determine species used for food; and we can recognize bones made into tools and jewelry (Hargrave 1965). Collectively bones provide much material for use by aboriginal peoples of the past and much intangible data for use by later man.

One aspect of highest importance, long overlooked, is the possible value of a bone for its own sake. Once removed from its associations in time and culture, discarded, or destroyed, the worth of faunal specimens can never be restored. Few of us realize that there are only three major land sources of biological remains, those from modern lifeforms at the top of the time scale, remains of lifeforms in paleohorizons toward the bottom of the scale, and those accumulated remains found in refuse heaps of human origin of intermediate horizons.

Bird or other bones may be found in almost any situation during the excavation of an abandoned human habitation or other place of use. The recovery of bones is incidental to specified investigations for other purposes. In any case, a defined area of study, properly called a "Provenience," is assigned a letter, number, symbol, or any combination thereof, which designation at all times identifies collectively all items from the provenience. Individual numbers, selected and assigned on a prearranged plan, identify separate items or groups of related material, as the case of closely associated head and wing bones of a bird would indicate skeletal remains of a stuffed birdskin (Fig. 23); the closely associated body bones of a bird could indicate that a birdskin had been prepared and the flesh-covered body had been discarded; the presence of several bones of a wing, not the humerus (Fig. 22), could indicate that a wing was used as paraphernalia or for an ornament. Attention to the relationship of elements being uncovered can result in the recognition of bone remains of once rare, feathered objects.

It is unlikely that any bones from artifacts would be altered by cutting. I have not found even one macaw bone that was humanly altered. These and other bird bones seemingly related should be packaged together and duly recorded.

Bird burials should be treated in like manner and care should be taken to collect some tracheal rings (windpipe), since I consider them to be proof that the bird was buried in the flesh. In like manner, the presence of calcified tendinal splints of the leg muscles of turkeys also indicate that the turkey was buried in the flesh (Hargrave 1965).

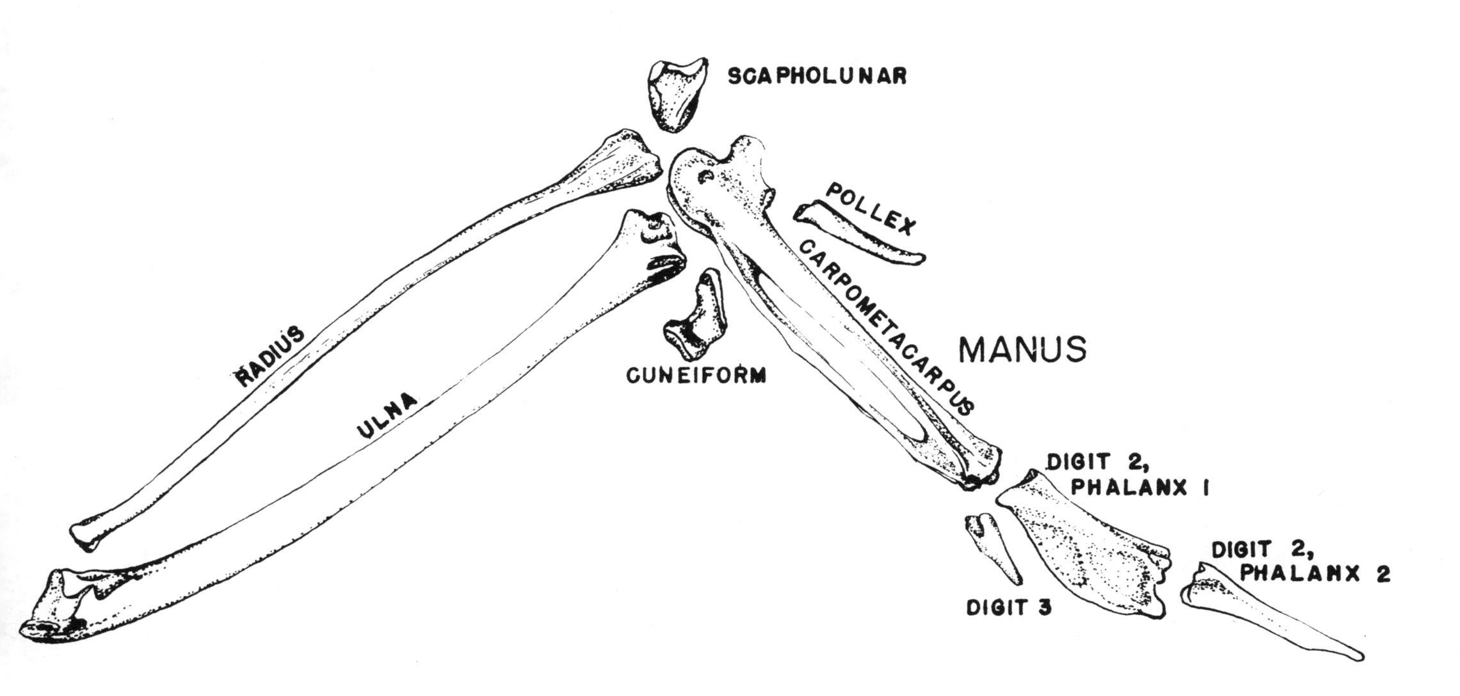

Fig. 22. Bones in position of articulation as they would appear in separate wing with feathers. (Scale: X 1)

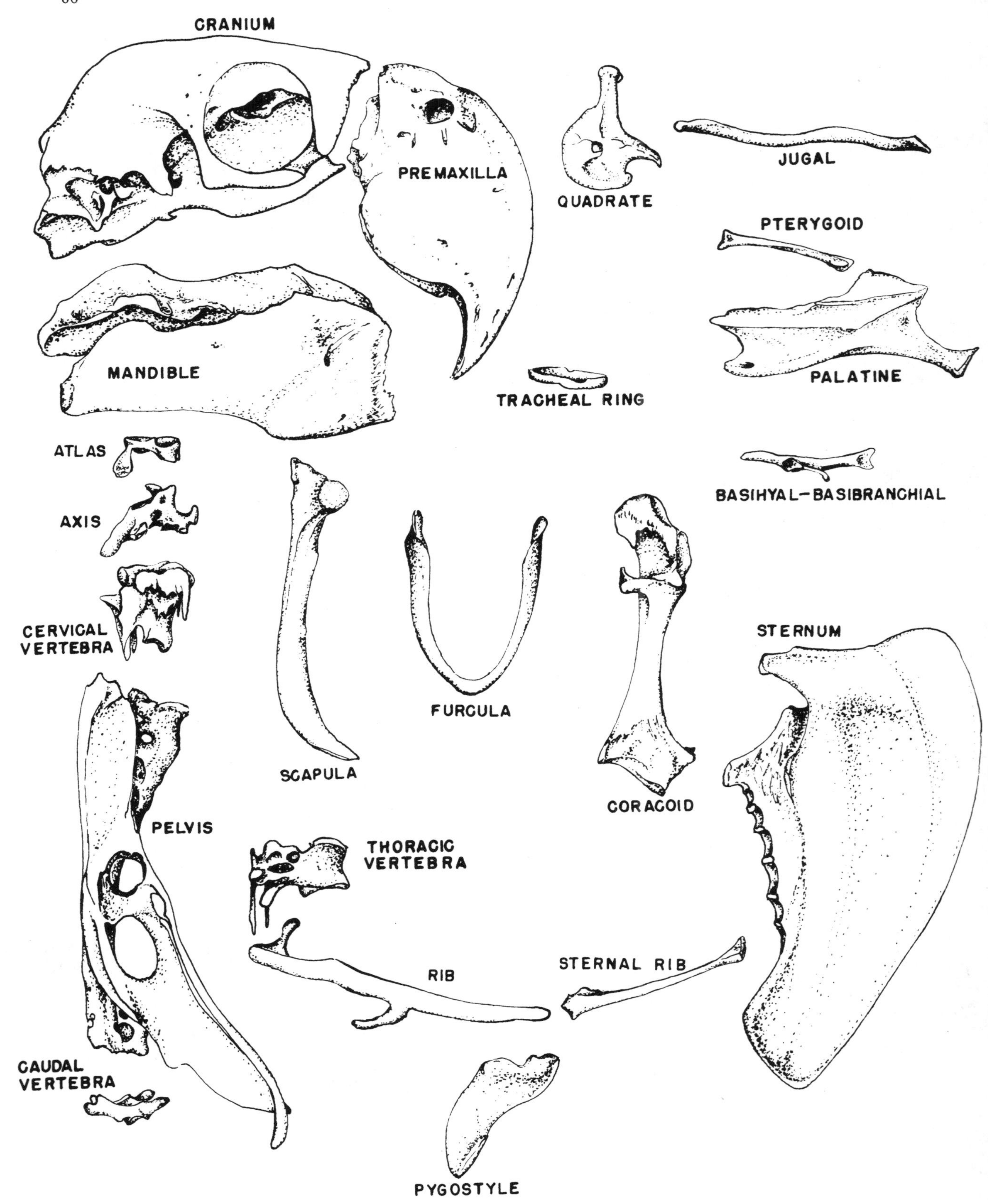

Fig. 23. Elements of the macaw skeleton. Prehistoric stuffed skins normally included head and lower wings. A single pygostyle might indicate a separate tail fan of feathers. (Scale: × 1)

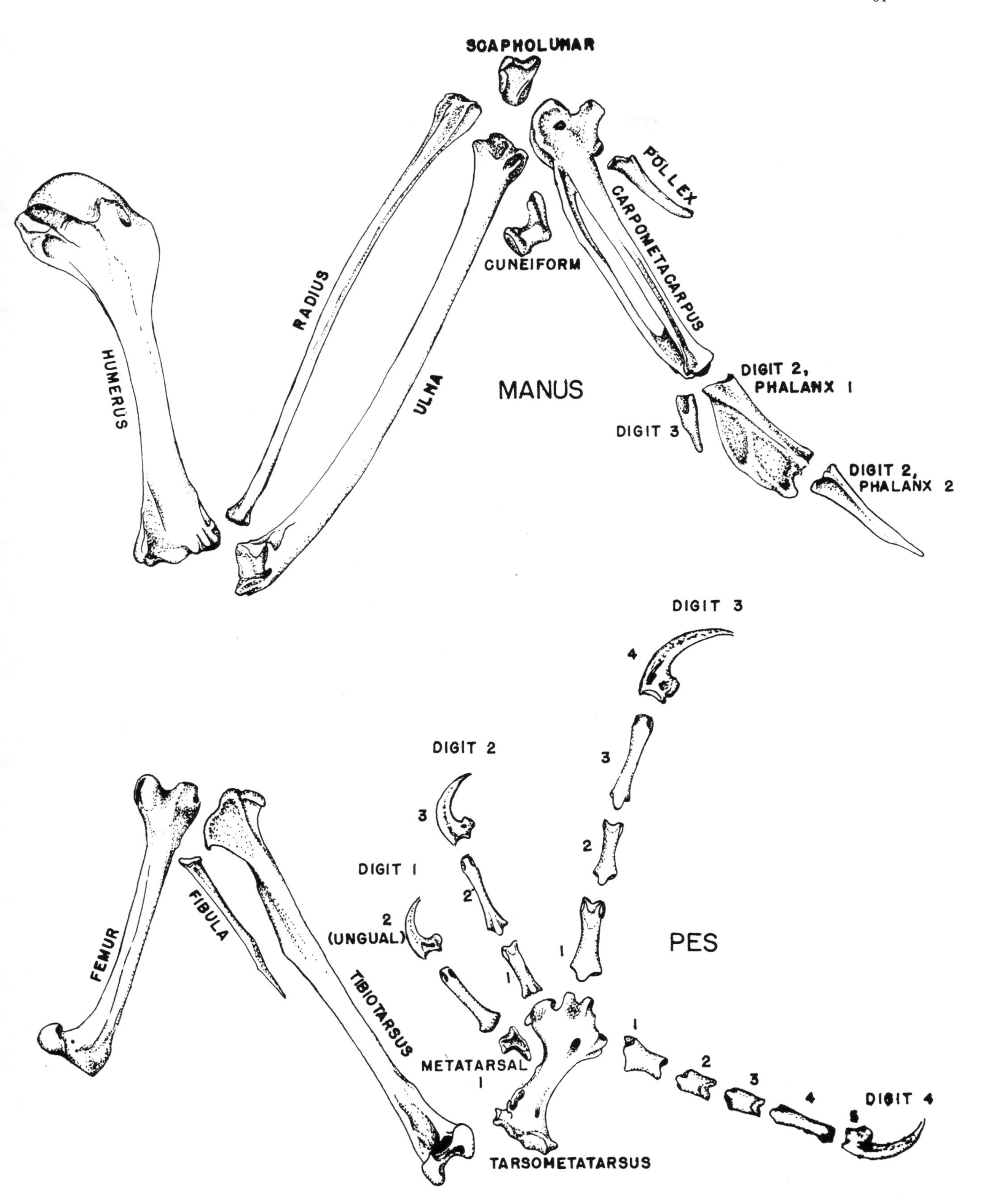
SCAPHOLUNAR
POLLEX
CARPOMETACARPUS
CUNEIFORM
RADIUS
HUMERUS
ULNA
MANUS
DIGIT 2,
PHALANX 1
DIGIT 3
DIGIT 2,
PHALANX 2
DIGIT 3
4
3
2
1
DIGIT 2
3
2
1
DIGIT 1
2
(UNGUAL)
FEMUR
FIBULA
TIBIOTARSUS
METATARSAL
1
TARSOMETATARSUS
PES
1
2
3
4
5
DIGIT 4

Appendix C

Care of Bird Bones in Field and Laboratory

To all archaeological field workers, applied ornithology frequently is limited to care of the material in the field. The handling of bird bones recovered at the excavation site is more a matter of caution, care, and common sense than it is knowledge of plaster and gauze and ways of wrapping and boxing. This is true of most archaeological sites in the Southwest, where bone refuse is usually dry or nearly so because of the characteristically dry soil of the region.

The important thing is to remove the bones in their entirety, if possible, and to protect them while at the site in such a manner that they can be transported safely. Freshly broken bones should be wrapped together, and all other closely associated bones should be packaged as a unit. Initial caution and care in treatment, however, should not be superseded by later neglect and carelessness. One should never throw bones onto a pile of anything or into a bag; one should never throw sherds or rocks, as a mano, onto a bone; one should never roughly handle bags or boxes filled with bones. Broken bones should be mended. Good treatment is only sensible; a little considerate care of the material and a little more appreciation of its value will be rewarding.

In every case a bag, a box, or a singly tagged bone should have a slip of paper with complete provenience data, site, name, or number placed inside the bag or box; the same provenience data should be written with a heavy soft lead pencil on the outside.

When material is received at a laboratory, processing should begin by lightly washing each individual bone without soaking, then drying it in the shade. When dried, all bones from the same provenience may be packaged together for further processing. In this state, bone collections of any size may be stored indefinitely.

If to be studied soon, on arrival at the base laboratory, material from one provenience designation at a time is laid out for study, since each bone is associated in time and culture with each object having the identical provenience designation. In this manner birds are segregated and packaged to be further analyzed according to groups, such as hawks, turkeys, or macaws. Species of a group can be separated and subjected to further study by age and size characters to determine the approximate number of individuals from the same provenience. After each individual is identified and defined by elements present, a catalogue number is assigned. Now, and not before, a reference number (which is the catalogue number) is inked in small letters and numerals on each bone of a practical size. Small uncatalogued bones are put in catalogued vials or boxes; these are filed with the large bones of the individual bird. It has been proved an unsound laboratory practice to mar the surface of a bone more than necessary for recording the catalogue number of the specimen. A heavily inked bone surface can be so distorted in appearance that much time may be required to clean the bone of ink for an unobstructed view of small characters which are sometimes essential to correct identification. I have found it quite practical to use transparent plastic bags for filing most bird bones, both during stages of study and after species identification has been made.

The permanent record for these bone specimens is a set of three catalogue cards with the same data, one for a catalogue number file, one for a species file, and one for a site file that is indexed by provenience and species. These comments on the care of bird bones apply to bones of any kind of animal.

GLOSSARY

Acetabulum. The cup-shaped socket in the pelvis, which receives the proximal head of the femur.

Anterior articular ligament (attachment on humerus). Depression on the interior margin of the distal end, palmar view, which is often the site of pathological accretions.

Antitrochanter (of pelvis). "An articular surface on the ilium of birds against which the great trochanter of the femur plays" (Websters New International Dictionary, 2nd Ed., Unabridged, 1954: p. 120).

Atlas. The first cervical vertebra, which articulates with the skull.

Axis. The second cervical vertebra, upon which the atlas rotates.

Basihyal-basibranchial. A composite bone providing the main support for the tongue among parrots.

Basitemporal plate. Among parrots, a flattened area on the ventral surface of the cranium immediately anterior to the foramen magnum. (Figs. 4*b*, 10*a*, and 10*b*)

Bicipital Crest (of humerus). Ventral margin of the proximal head of the humerus, curving from the articulation with the coracoid to the humeral shaft. (Fig. 13*e*)

Carinal. Referring to the keel of the sternum.

Carpometacarpus. A compound wing bone formed from distal carpal and metatarpal bones. (Fig. 23)

Caudal. Referring to the tail.

Cervical. Referring to the neck.

Coracoid. A bone of the pectoral girdle, articulating ventrally with the sternum, and dorsally with the furcula, scapula and humerus. (Fig. 23)

Cranial. Referring to the skull.

Cranium. The skull.

Cuneiform. The posterior of two carpal bones of the wing. (Fig. 23)

Eustachian tubes. "A long and cartilaginous tube connecting the . . . middle ear with the nasopharynx and serving to equalize air pressure on both sides of the tympanic membrane." (Websters New International Dictionary, 2nd Ed., Unabridged 1954: p. 882). Its opening in the ventral surface is often the site of pathology, perhaps as a result of respiratory infections or fungal afflictions. (Fig. 6 *d*)

Fenestra (of cranium). A windowlike opening. (Fig. 6*d*)

Femur. Proximal bone of the leg. (Fig. 23)

Fibula. An incomplete bone of the lower leg parallel and lateral to the proximal portion of the tibiotarsus. (Fig. 23)

Foramen Magnum. "The large opening in the occipital bone through which the spinal cord passes . . . ." (Websters New International Dictionary, 2nd Ed., Unabridged, 1954: p. 985).

Furcula. The fused clavicles, "wishbone." (Fig. 23)

Humerus. The bone of the upper wing. (Fig. 23)

Inflated. Swollen, distended.

Jugal. A facial bone, which, in parrots, articulates with the quadrate and the premaxilla.

Juvenal. First true plumage containing contour feathers after loss of natal down.

Juvenile. A young, immature bird.

Mandible. Lower half of bill. (Fig. 23)

Manubrium. A process of the anterior margin of the sternum which lies at the base of the keel.

Manus. The hand, including carpometacarpus and phalanges.

Palatine. A pair of bones lying in the roof of the mouth, which among parrots, join the mid-portion of the premaxilla to the ventral surface of the cranium. (Fig. 23)

Pelvis. The pelvic girdle, composed of ilium, pubis, and synsacrum. (Fig. 23)

Premaxilla. The upper half of the beak. (Fig. 23)

Pes. The foot, including tarsometatarsus, metatarsal I, and phalanges. (Fig. 23)

Pollex (thumb). First digit of the forelimb. (Fig. 23)

Postorbital process. Posterior member of suborbital arch. (Fig. 6*a*)

Preorbital process. Anterior member of suborbital arch, which grows in an arc until it overlaps medially the postorbital process. (Fig. 6*a*)

Pterygoid. A slender bone articulating with the palatine and the quadrate. (Fig. 23)

Pygostyle. The tail bone, formed by the fusing of the terminal caudal vertebrae.

Quadrate. A three pronged bone which hinges the mandible to the cranium and to the pterygoid bone. (Fig. 23)

Radius. The more slender bone of the forewing. (Fig. 23)

Rib and sternal rib. A series of paired curved bones surrounding the thoracic cavity. Ribs are often branched and articulate with the vertebrae. Sternal ribs are short, unbranched, and articulate with the sternum. (Fig. 23)

Roughened ulna. A pathological condition of the ulna characterized by diagonal ridges on the anconal surface of the shaft. (Fig. 17*b*)

Scapholunar. The anterior of two carpal bones of the wing. (Fig. 23)

Scapula. The shoulder blade. (Fig. 23)

Sternum. The breastbone. (Fig. 23)

Suborbital arch. An arch formed by an extension of the preorbital process which grows along the lower margin of the orbit and then upward to overlap and fuse with the medial surface of the postorbital process at an age of ca. 11 months in macaws.

Synsacrum. The region of the vertebral column included in the pelvis. (Fig. 2*a*)

Tarsometatarsus. The compound bone of the avian foot which bears the toes. (Fig. 23)

Taxon (pl. taxa). A taxonomic unit or category.

Tibiotarsus. The larger bone of the lower leg, the "drumstick." (Fig. 23)

Ulna. The larger bone of the forewing to which the secondary flight feathers are attached. (Fig. 23)

Vertebrae (s. vertebra). The vertebral column or backbone through which passes the spinal cord. (Fig. 23)

REFERENCES

AMERICAN ORNITHOLOGISTS' UNION
1957 *Check-List of North American Birds,* Fifth Edition. Lord Baltimore Press, Baltimore.

AUSTIN, OLIVER L., JR.
1961 *Birds of the World.* Golden Press, New York.

BANDELIER, ADOLPH F.
1890 Contributions to the History of the Southwestern Portion of the United States. *Papers of the Archaeological Institute of America, American Series,* No. 5. Peabody Museum of American Archaeology and Ethnology, Harvard University, Cambridge.

BANNISTER, BRYANT
1965 Tree-Ring Dating of the Archeological Sites in the Chaco Region, New Mexico. *Southwestern Monuments Association, Technical Series,* Vol. 6, Part 2. Globe.

BANNISTER, BRYANT, ELIZABETH A. M. GELL, AND JOHN W. HANNAH
1966 *Tree-Ring Dates from Arizona N – Q; Verde, Show Low, St. Johns Area.* Laboratory of Tree-Ring Research, University of Arizona, Tucson.

BANNISTER, BRYANT, JOHN W. HANNAH, AND WILLIAM J. ROBINSON.
1966 *Tree-Ring Dates from Arizona K: Puerco, Wide Ruin, Ganado Area.* Laboratory of Tree-Ring Research, University of Arizona, Tucson.

BEALS, RALPH L., GEORGE W. BRAINARD, AND WATSON SMITH.
1945 Archaeological Studies in Northeast Arizona. *University of California Publications in American Archaeology and Ethnology,* Vol. 44, No. 1. University of California Press, Berkeley and Los Angeles.

BLAKE, EMMET REID
1953 *Birds of Mexico.* University of Chicago Press, Chicago.

BOLTON, HERBERT E., EDITOR
1916 *Spanish Explorations in the Southwest, 1542 to 1706.* Charles Scribner's Sons, New York.

BRADFIELD, WESLEY
1929 Cameron Creek Village, A Site in the Mimbres Area in Grant County, New Mexico. *Monographs of the School of American Research,* No. 1. Santa Fe.

CAYWOOD, LOUIS R., AND EDWARD H. SPICER
1935 *Tuzigoot, the Excavation and Repair of a Ruin on the Verde River Near Clarkdale, Arizona.* National Park Service, Berkeley.

COLTON, HAROLD S.
1946 The Sinagua. A Summary of the Archaeology of the Region of Flagstaff, Arizona. *Museum of Northern Arizona, Bulletin 22.* Flagstaff.

COLTON, H. S., AND LYNDON L. HARGRAVE
1937 Handbook of Northern Arizona Pottery Wares. *Museum of Northern Arizona, Bulletin* 11. Flagstaff.

CUMMINGS, BYRON
1940 *Kinishba, A Prehistoric Pueblo of the Great Pueblo Period.* Hohokam Museums Association and the University of Arizona, Tucson.

DEAN, JEFFREY S.
1964 *Final Report; Kiet Siel Tree-Ring Dating Project.* Laboratory of Tree-Ring Research, University of Arizona, Tucson.

DELACOUR, JEAN
1964 *The Waterfowl of the World,* Vol. 4. Country Life, London.

DE SCHAUENSEE, R. MEYER
1964 *The Birds of Colombia.* Livingston, Narberth (Pennsylvania).

DI PESO, CHARLES C.
1958 The Reeve Ruin of Southeastern Arizona. *The Amerind Foundation,* No. 8. Dragoon.

DOUGLASS, ANDREW ELLICOTT
1929 The Secret of the Southwest Solved by Talkative Tree Rings. *National Geographic Magazine*, Vol. 56, No. 6, pp. 737-70. Washington.

FRIEDMANN, HERBERT, LUDLOW GRISCOM, AND ROBERT T. MOORE
1950 Distributional Check-List of the Birds of Mexico, Part I. *Cooper Ornithological Club, Pacific Coast Avifauna,* No. 29. Berkeley.

HARGRAVE, LYNDON L.
1931a First Mesa. *Museum Notes,* Vol. 3, No. 8. Museum of Northern Arizona, Flagstaff.
1931b Excavations at Kin Tiel and Kokopnyama. *Smithsonian Miscellaneous Collections,* Vol. 82, No. 11, pp. 80-120. Washington.
1932 Guide to Forty Pottery Types from the Hopi Country and the San Francisco Mountains, Arizona. *Museum of Northern Arizona Bulletin* 1. Flagstaff.
1933 The Museum of Northern Arizona Archeological Expedition, 1933; Wupatki National Monument. *Museum Notes,* Vol. 6, No. 5. Museum of Northern Arizona, Flagstaff.
1934 The Tsegi Country. *Museum Notes,* Vol. 6, No. 11. Museum of Northern Arizona, Flagstaff.
1935 Archaeological Investigations in the Tsegi Canyons of Northeastern Arizona in 1934. *Museum Notes,* Vol. 7, No. 7. Museum of Northern Arizona, Flagstaff.
1937 Sikyatki; Were the Inhabitants Hopi? *Museum Notes,* Vol. 9, No. 12. Museum of Northern Arizona, Flagstaff.
1938 A Plea for More Careful Preservation of All Biological Material from Prehistoric Sites. *Southwestern Lore,* Vol. 4, No. 3, p. 47. Gunnison.
1965 Turkey Bones from Wetherill Mesa. In "Contributions of the Wetherill Mesa Archeological Project," assembled by Douglas Osborne, pp. 161-6. *Memoirs of the Society for American Archaeology,* No. 19. Salt Lake City.
1968 The Proper Way to Report Birds in Print. *American Antiquity,* Vol. 33, No. 3, pp. 384-5. Salt Lake City.

HAURY, EMIL W.
1936 Some Southwestern Pottery Types, Series IV. *Medallion Papers,* No. 19. Gila Pueblo, Globe.
1958 Evidence at Point of Pines for a Prehistoric Migration from Northern Arizona. In "Migrations in New World Culture History," edited by Raymond H. Thompson, pp. 1-6. *University of Arizona Bulletin,* Vol. 29, No. 2, *Social Science Bulletin* 27. Tucson.

HEWETT, EDGAR LEE
1936 *The Chaco Canyon and Its Monuments.* University of New Mexico Press, Albuquerque.

JACKSON, EARL, AND SALLIE PIERCE VAN VALKENBURGH
1954 Montezuma Castle Archeology; Part I; Excavations. *Southwestern Monuments Association, Technical Series,* Vol. 3, No. 1. Globe.

JOHNSON, ALFRED E.
1965 *The Development of the Western Pueblo Culture.* Doctoral dissertation, University of Arizona. University Microfilms, Ann Arbor.

JUDD, NEIL M.
1954 The Material Culture of Pueblo Bonito. *Smithsonian Miscellaneous Collections,* Vol. 124. Washington.
1959 Pueblo del Arroyo, Chaco Canyon, New Mexico. *Smithsonian Miscellaneous Collections,* Vol. 138, No. 1. Washington.

KIDDER, ALFRED VINCENT
1924 *An Introduction to the Study of Southwestern Archaeology.* Yale University Press, New Haven.
1931 *The Pottery of Pecos, Volume I.* Yale University Press, New Haven.

KING, DALE S.
1949 Nalakihu, Excavations at a Pueblo III Site on Wupatki National Monument, Arizona. *Museum of Northern Arizona, Bulletin* 23. Flagstaff.

LOCKETT, H. CLAIRBORNE, AND LYNDON L. HARGRAVE
1953 Woodchuck Cave, A Basket Maker II Site in Tsegi Canyon, Arizona. *Museum of Northern Arizona, Bulletin* 26. Flagstaff.

McGregor, John C.
1941 Winona and Ridge Ruin, Part I: Architecture and Material Culture. *Museum of Northern Arizona, Bulletin* 18. Flagstaff.
1943 Burial of an Early American Magician. *Proceedings of the American Philosophical Society*, Vol. 86, No. 2, pp. 270-98. Philadelphia.

Nesbitt, Paul H.
1931 The Ancient Mimbreños, Based on Investigations at the Mattocks Ruin, Mimbres Valley, New Mexico. *The Logan Museum, Bulletin* No. 4. Beloit College, Beloit.

Olsen, Stanley J.
1967 Osteology of the Macaw and Thick-billed Parrot. *Kiva,* Vol. 32, No. 3, pp. 57-72. Tucson.

Pepper, George H.
1920 Pueblo Bonito. *Anthropological Papers of the American Museum of Natural History*, Vol. 27. New York.

Peters, James Lee
1937 *Check-List of Birds of the World*, Vol. 3. Harvard University Press, Cambridge.

Robinson, William J.
1958 A New Type of Ceremonial Pottery Killing at Point of Pines. *Kiva,* Vol. 23, No. 3, pp. 12-14. Tucson.

Robinson, William J., and Roderick Sprague
1965 Disposal of the Dead at Point of Pines, Arizona. *American Antiquity*, Vol. 30, No. 4, pp. 442-53. Salt Lake City.

Smiley, Terah L.
1951 A Summary of Tree-Ring Dates from some Southwestern Archaeological Sites. *University of Arizona Bulletin*, Vol. 22, No. 4, *Laboratory of Tree-Ring Research Bulletin*, No. 5. Tucson.

Smith, Watson
1952 Kiva Mural Decorations at Awatovi and Kawaika-a. *Papers of the Peabody Museum of American Archaeology and Ethnology, Harvard University,* Vol. 37. Cambridge.

Thompson, D'Arcy W.
1899 On Characteristic Points in the Cranial Osteology of the Parrots. *Proceedings of the Zoological Society of London*, London.

Thompson, Raymond H., and William A. Longacre
1966 The University of Arizona Archaeological Field School at Grasshopper, East Central Arizona. *Kiva*, Vol. 31, No. 4, pp. 255-75. Tucson.

Van Rossem, A. J.
1945 A Distributional Survey of the Birds of Sonora, Mexico. *Louisiana State University, Museum of Zoology, Occasional Papers*, No. 21. Louisiana State University Press, Baton Rouge.

Vivian, Gordon
1964 Gran Quivira, Excavations in a 17th Century Jumano Pueblo. *Archeological Research Series,* No. 8. National Park Service, Washington.

Vivian, Gordon, and Tom W. Mathews
1965 Kin Kletso: A Pueblo III Community in Chaco Canyon, New Mexico. *Southwestern Monuments Association, Technical Series,* Vol. 6, Part I. Globe.

Wasley, William W.
1960 A Hohokam Platform Mound at the Gatlin Site, Gila Bend, Arizona. *American Antiquity,* Vol. 26, No. 2, pp. 244-62. Salt Lake City.

Wetherill, John
1934 Navajo National Monument. *The Southwestern Monuments Monthly Report*, January - April, 1934. Coolidge.

Woodbury, Richard B.
1961 Prehistoric Agriculture at Point of Pines, Arizona. *Memoirs of the Society for American Archaeology*, No. 17. Salt Lake City.

Wyllys, Rufus Kay
1931 Padre Luis Velarde's Relacion of Pimeria Alta, 1716. *New Mexico Historical Review*, Vol. 6, No. 2, p. 111. Albuquerque.